Safety on the Rig

Fourth Edition, Revised

By William E. Jackson

Published by

PETROLEUM EXTENSION SERVICE
THE UNIVERSITY OF TEXAS AT AUSTIN
Division of Continuing & Innovative Education
Austin, Texas

Originally produced by

INTERNATIONAL ASSOCIATION
OF DRILLING CONTRACTORS
Houston, Texas

2014

Library of Congress Cataloging-in-Publication Data

Jackson, William E., 1929—
    Safety on the rig / written by William E. Jackson. — 4th ed.
      p. cm. — (Rotary drilling series ; unit I, lesson 10)
      ISBN 0-88698-186-7 (pbk. : alk. paper)
     1. Oil field—Safety measures. I. Title. III. Series
    TN871.J33 1999
    622'.8—dc21

                         2003020668
                              CIP

Disclaimer

Although all reasonable care has been taken in preparing this publication, the authors, the Petroleum Extension Service (PETEX™) of The University of Texas at Austin, and any other individuals and their affiliated groups involved in preparing this content, assume no responsibility for the consequences of its use. Each recipient should ensure he or she is properly trained and informed about the unique policies and practices regarding application of the information contained herein. Any recommendations, descriptions, and methods in this book are presented solely for educational purposes.

Catalog no. 2.110401
ISBN 0-88698-186-7
    978-0-88698-186-0

No state tax funds were used to publish this book. The University of Texas at Austin is an equal opportunity employer.

Copy Editor: Virginia Dosher

Graphic Designer: Debbie Caples

Contents

Figures

Tables

Foreword

For many years, PETEX's Rotary Drilling Series has oriented new personnel and assisted experienced hands in the rotary drilling industry. The industry has changed greatly since the publication of the first edition of *Safety on the Rig* in 1967, and the manual has been updated to reflect those changes. This revised fourth edition addresses topics such as safety offshore and the importance of HSE—health, safety, and the environment—on today's rigs, where safety comes first.

This revised edition of *Safety on the Rig* has been updated with current information on the topic. The layout has been made more open, with each page or two presenting a topic that is easy to digest and find again later. Notes throughout the chapters in the margins and summaries at the end of the chapters call out important points for the reader. The glossary defines important words used throughout the book. In addition, large, full-color illustrations aid in understanding the material, and a self-test at the end of the book serves as a review or a way for the reader to assess his or her knowledge.

PETEX™ (Petroleum Extension Service)
The University of Texas at Austin

Preface

A drilling rig is a dangerous work environment. Every year, more than 100 fatalities occur in the oil and gas extraction industry in the United States; many more times, permanent injuries occur. There are high human and financial costs to constructing rigs and pipelines, drilling wells, extracting gas or oil, and transporting fluids to refineries.

Many accidents, however, are avoidable. The purpose of *Safety on the Rig*, Fourth Edition, Revised, is to prevent accidents on the rig or, when accidents do occur, to instruct crew members in measures to reduce the harm that results from them.

The book is an introductory-level survey of the topic, addressing precautions for the most dangerous activities on the rig, personal safety equipment, fire prevention and suppression, the handling of hazardous substances, and emergency first aid. It is also useful for seasoned rig hands who are looking to remember important topics from their training.

Every rig crewmember must be thoroughly trained in safe practices. Safety on the Rig provides information on safe practices that can be part of a rigorous training program.

PETEX™ (Petroleum Extension Service)
The University of Texas at Austin

Acknowledgments

PETEX would like to sincerely thank the many people who contributed to previous editions of *Safety on the Rig*, which formed the basis of this Fourth Edition, Revised.

The following people graciously contributed background information and reviewed manuscript.

Ben Woodward

Harry Olds

Robert Wilson

Linda Guthrie

Fed Calhoun

John Thomas

We would also like to express our appreciation to the American Red Cross and the International Association of Drilling Contractors (IADC) for permission to adapt original illustrations and content from their publications.

In addition, we would like to express gratitude to industry reviewers and readers for their invaluable assistance in the revision of the Rotary Drilling series.

On the PETEX staff, Kathryn Roberts originally saw production through from idea to printed book, and Doris Dickey proofread innumerable versions of the manuscript. For this edition, Debbie Caples designed the layout, and Virginia Dosher served as copy editor.

Units of Measurement

Throughout the world, two systems of measurement dominate: the English system and the metric system. Today, the United States is one of only a few countries that employ the English system.

The English system uses the pound as the unit of weight, the foot as the unit of length, and the gallon as the unit of capacity. In the English system, for example, 1 foot equals 12 inches, 1 yard equals 36 inches, and 1 mile equals 5,280 feet or 1,760 yards.

The metric system uses the gram as the unit of weight, the metre as the unit of length, and the litre as the unit of capacity. In the metric system, 1 metre equals 10 decimetres, 100 centimetres, or 1,000 millimetres. A kilometre equals 1,000 metres. The metric system, unlike the English system, uses a base of 10; thus, it is easy to convert from one unit to another. To convert from one unit to another in the English system, you must memorize or look up the values.

In the late 1970s, the Eleventh General Conference on Weights and Measures described and adopted the Système International (SI) d'Unités. Conference participants based the SI system on the metric system and designed it as an international standard of measurement.

The Rotary Drilling Series gives both English and SI units. And because the SI system employs the British spelling of many of the terms, the book follows those spelling rules as well. The unit of length, for example, is metre, not meter. (Note, however, that the unit of weight is gram, not gramme.)

To aid U.S. readers in making and understanding the conversion system, we include the table on the next page.

English-Units-to-SI-Units Conversion Factors

Quantity or Property	English Units	Multiply English Units By	To Obtain These SI Units
Length, depth, or height	inches (in.)	25.4	millimetres (mm)
		2.54	centimetres (cm)
	feet (ft)	0.3048	metres (m)
	yards (yd)	0.9144	metres (m)
	miles (mi)	1609.344	metres (m)
		1.61	kilometres (km)
Hole and pipe diameters, bit size	inches (in.)	25.4	millimetres (mm)
Drilling rate	feet per hour (ft/h)	0.3048	metres per hour (m/h)
Weight on bit	pounds (lb)	0.445	decanewtons (dN)
Nozzle size	32nds of an inch	0.8	millimetres (mm)
Volume	barrels (bbl)	0.159	cubic metres (m^3)
		159	litres (L)
	gallons per stroke (gal/stroke)	0.00379	cubic metres per stroke (m^3/stroke)
	ounces (oz)	29.57	millilitres (mL)
	cubic inches (in.^{3})	16.387	cubic centimetres (cm^3)
	cubic feet (ft^3)	28.3169	litres (L)
		0.0283	cubic metres (m^3)
	quarts (qt)	0.9464	litres (L)
	gallons (gal)	3.7854	litres (L)
	gallons (gal)	0.00379	cubic metres (m^3)
	pounds per barrel (lb/bbl)	2.895	kilograms per cubic metre (kg/m^3)
	barrels per ton (bbl/tn)	0.175	cubic metres per tonne (m^3/t)
Pump output and flow rate	gallons per minute (gpm)	0.00379	cubic metres per minute (m^3/min)
	gallons per hour (gph)	0.00379	cubic metres per hour (m^3/h)
	barrels per stroke (bbl/stroke)	0.159	cubic metres per stroke (m^3/stroke)
	barrels per minute (bbl/min)	0.159	cubic metres per minute (m^3/min)
Pressure	pounds per square inch (psi)	6.895	kilopascals (kPa)
		0.006895	megapascals (MPa)
Temperature	degrees Fahrenheit (°F)	$\dfrac{°F - 32}{1.8}$	degrees Celsius (°C)
Mass (weight)	ounces (oz)	28.35	grams (g)
	pounds (lb)	453.59	grams (g)
		0.4536	kilograms (kg)
	tons (tn)	0.9072	tonnes (t)
	pounds per foot (lb/ft)	1.488	kilograms per metre (kg/m)
Mud weight	pounds per gallon (ppg)	119.82	kilograms per cubic metre (kg/m^3)
	pounds per cubic foot (lb/ft^3)	16.0	kilograms per cubic metre (kg/m^3)
Pressure gradient	pounds per square inch per foot (psi/ft)	22.621	kilopascals per metre (kPa/m)
Funnel viscosity	seconds per quart (s/qt)	1.057	seconds per litre (s/L)
Yield point	pounds per 100 square feet (lb/100 ft^2)	0.48	pascals (Pa)
Gel strength	pounds per 100 square feet (lb/100 ft^2)	0.48	pascals (Pa)
Filter cake thickness	32nds of an inch	0.8	millimetres (mm)
Power	horsepower (hp)	0.75	kilowatts (kW)
Area	square inches (in.^{2})	6.45	square centimetres (cm^2)
	square feet (ft^2)	0.0929	square metres (m^2)
	square yards (yd^2)	0.8361	square metres (m^2)
	square miles (mi^2)	2.59	square kilometres (km^2)
	acre (ac)	0.40	hectare (ha)
Drilling line wear	ton-miles (tn•mi)	14.317	megajoules (MJ)
		1.459	tonne-kilometres (t•km)
Torque	foot-pounds (ft•lb)	1.3558	newton metres (N•m)

Introduction

In this chapter:

- The hazardous work environment of the drilling rig
- The most dangerous activities for those working in the oil and gas extraction industry
- The importance of complying with OSHA safety regulations

Working on a drilling rig is a tough, hazardous job where safety is a critical issue. A striking statistic illustrates that point: Between 2003 and 2012, there were 1,077 fatalities of oil and gas workers at U.S. job sites (fig. 1). To have an idea of how significant this number is, consider the following:

- There were approximately 450,000 workers in the industry in 2011.
- The fatality rate in the industry is several times higher than the overall rate for all U.S. industries.

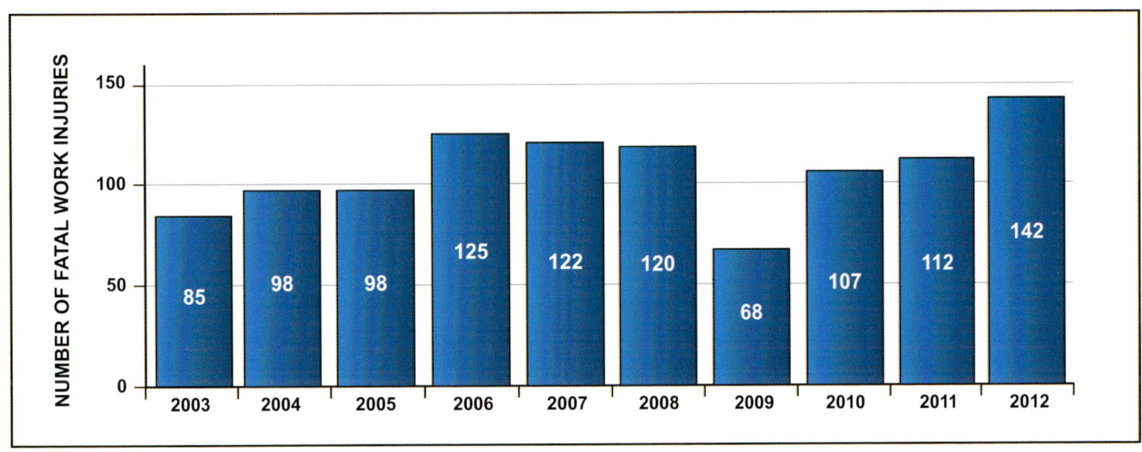

Figure 1.   Fatalities in the U.S. oil and gas extraction industry number, on average, over 100 per year.

1

Government regulations that improve safety practices in the U.S. workplace are enforced by OSHA, the Occupational Safety and Health Administration.

Oil and gas workers are involved in a large variety of potentially dangerous tasks. For example, they operate heavy machinery that moves in restricted places on the rig floor; they mix chemicals, handle fuels, and repair electrical equipment; and they work several stories off the ground. They also work in settings where weather can suddenly become severe and cause equipment failures or where well conditions can change rapidly, causing gas leaks or high pressures.

Dangers exist in almost all job functions. Figure 2 shows the causes of fatal injuries at oil and gas work sites in the U.S. between 2003 and 2009. An examination of the chart reveals the top five causes of injury:

- Highway crashes
- Being struck by an object
- Explosions
- Being caught in moving machinery
- Falling

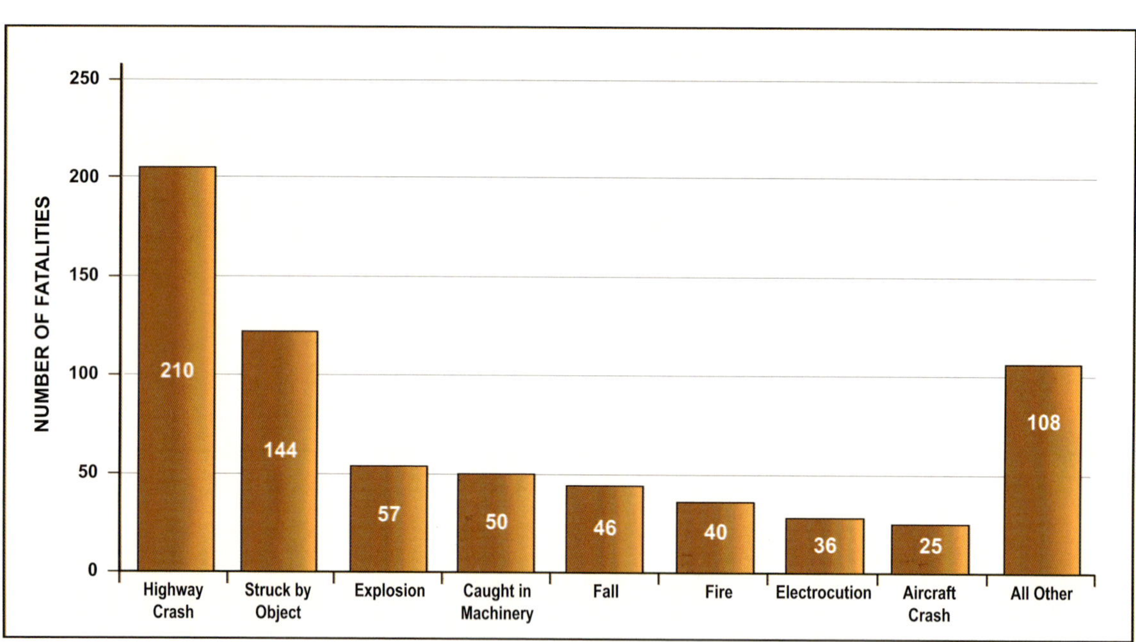

Figure 2. Most common fatal events in the U.S. oil and gas extraction industry from 2003 to 2009

Rigorous safety training of all personnel in all phases of operations is necessary; safe procedures must be learned and practiced. It is every person's responsibility to be alert and to implement training at all times (fig. 3).

When accidents happen, there can be great risks to personnel. A careless act by one person can cause the hole, the rig, or all the valuable oil and gas to be lost. These results are serious but do not compare to the danger to the crewmembers, who are at risk to lose a limb, an eye, or their lives.

The Occupational Safety and Health (OSH) Act of 1970 is the set of federal laws that govern health and safety practices affecting workers. The Occupational Safety and Health Administration (OSHA) is the federal agency that interprets the general laws to set specific regulations for each industry; it also enforces the regulations in the workplace.

OSHA has greatly affected practices in the drilling and oilfield services industries. The regulations are numerous and complex. They aim to protect employees by, for example, limiting their exposure to chemicals, requiring them to wear harnesses to prevent falls, and instructing them in proper ergonomic positions. To enforce the regulations, various state, regional, and national entities inspect oilfield work sites on a regional basis. Individual contractors have their own safety programs that incorporate government regulations.

Under OSHA regulations, some employees have to produce frequent reports or comply with other requirements that seem unnecessary and burdensome. Those employees should remember that compliance is required by the law and that it ensures a safer work place for all.

Figure 3. All workers must put safety first.

To summarize—

- Extracting oil and gas is hazardous work. Workplace accidents account for an average of more than 100 fatalities among U.S. oil and gas employees each year.

- The most common causes of injury are highway accidents, being struck by an object, explosions, being caught in moving machinery, and falling.

- There are risks for all employees, however.

- OSHA has set regulations to protect oilfield services workers. OSHA also enforces the regulations with the help of some other regulatory agencies.

- All employees should comply with OSHA regulations to improve safety in the workplace.

People and Safety

In this chapter:

- The chain of command for rig safety
- Responsibilities of individual crew members for safety
- How rig safety programs are developed
- The role of the safety director on a rig

Less than 5% of rig accidents are caused by mechanical failure. It is people who make rig operations safe, so everyone must develop a sense for safety and practice self-improvement. Before acting, an individual must automatically consider, "Will this put anyone in danger?" Every crewmember must be ever alert to risky or unsafe situations. Being alert is the best way to avoid injury to yourself and the crew. It is critical that a new crewmember receives supervision and instruction in safe operating procedures immediately upon reporting for work. Industry data show that 60% or more of rig injuries involve people on the job less than 6 months. Even an experienced person may need supervision, especially if coming from a different type or size of rig.

Operator

Responsibility for rig safety follows the corporate chain of command. All are involved, from the operator down to the new floorhand.

Depending on the contract, an operator may specify the casing and mud programs to be followed, furnish the casing, have it delivered, inspected, and specify the setting depth. If operators furnish the mud, they will control the type and weight and be responsible for having ample supplies on location or nearby to control the formation pressures encountered. Such matters are usually covered in the drilling contract. An operator must select a contractor with a rig capable of safely handling the job. The superintendent, field foreman, or consulting drilling engineer may be on location at crucial times and be in direct control of operations. At such times, the operator is effectively in charge of seeing that the work is performed without endangering the crew, equipment, or well.

Drilling Contractor or Rig Owner

Figure 4. Rig owner and safety engineer

The drilling contractor is primarily responsible for rig safety. The contractor devises a safety program after consulting with the rig owner.

A contractor has the primary responsibility for safe drilling operations. The rig owner determines the safety policy and instructs the various supervisors to implement that policy, showing a genuine interest and actively participating in developing all aspects of the company's safety programs. The rig's safety record should be as important as performance. A good contractor will never allow cost-cutting measures to compromise crew safety. The contractor has a responsibility to the rig crew and to the client to provide safe drilling equipment. All equipment should be properly sized to safely handle expected load limits, pressures, and depth. All needed safeguards should be in place and meet or exceed OSHA and other regulatory standards.

The contractor should be aware of and in compliance with all governmental regulations pertaining to drilling operations, including permit requirements for certain jobs performed by the crew. Responsibility for safety training is given to the drilling superintendent, rig managers, and drillers. They in turn ensure that all personnel under their supervision participate in safety training and education. To underscore their interest in safe operations, some contractors provide safety bonuses to be shared by crew and supervisors upon the safe completion of a well.

Most multi-rig contractors have a safety engineer or director charged with the overall responsibility for safety matters. Usually, safety personnel report directly to the company president or owner. Should an accident occur, they conduct a thorough investigation, prepare all reports, and develop procedures to avoid a reoccurrence of the incident. At the direction of the owner, the safety director may implement a job safety analysis (JSA) program enlisting support and input from all personnel in order to achieve a safe work site.

Drilling Superintendent

Drilling superintendents, or drilling managers, are in top management positions and usually report directly to the owner or company president. A superintendent may be in charge of the overall operations of several rigs, including the performance of the rig managers, drillers, crewmembers, subcontractors, service company vendors, and other field personnel. The superintendent is in contact with the operator's management both before and during drilling, and learns the capabilities of all the rigs and crews under supervision. Other aspects of the job may be to draw up the drilling contract or assure the client that the assigned rig can do the job safely and efficiently.

The superintendent works with the rig managers to see that the equipment is rigged up safely at the location. Unusual conditions at the location are dealt with properly. Rig managers are informed of any changes in safety procedures or equipment on the rig and are required to conduct safety training for the crew. Training will include periodic rig inspections meant to spot potentially hazardous conditions and ensure that such conditions are eliminated. Rig managers also deal with service companies, consultants, equipment engineers, and others working to improve rig performance and safety.

Rig Manager (Toolpusher)

The toolpusher's job has been redefined to include a broader range of responsibilities than it used to—hence the modern title "rig manager." The rig manager reports directly to the drilling superintendent (fig. 5). No longer is the main responsibility simply to get the hole drilled efficiently and quickly. The job has expanded to include direct control over safety, environmental, and other regulatory concerns in the field. The landowner's concerns over access and water must be addressed. Supply firms must be contacted and supervised. Visitors must be controlled and made aware of safety procedures.

The rig manager makes sure each driller keeps all operations within the rig's capability and that each driller trains the crew to work safely. Through experience, the hazards on the rig are known and eliminated promptly. If a job safety analysis program is in place, the rig manager has the primary role in its implementation. Periodic inspections are performed with each driller's participation and input. A rig safety list is filled out and the inspection results are reported to the superintendent. The manager investigates every accident, analyzes it with the driller and crew, and oversees needed corrective measures.

Figure 5. Drilling superintendent consulting with the rig manager

Driller

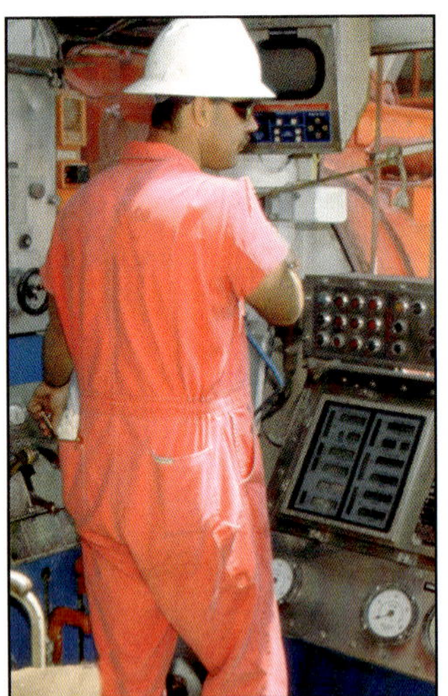

Figure 6. Driller at his position on the rig floor

The driller reports directly to the rig manager. The driller is in charge of the rig and crew on the location. From the driller's console area, he or she controls the drawworks and brakes, sets the bit weight, rotary speed, and pump pressure (fig. 6). Experience is important because errors in judgment can lead to damage to the rig or injury to the crew. The rig crew must trust the driller completely because safety depends a great deal on a driller's skill. A driller must be constantly aware of safety—not only of behavior and actions, but also of each person on the rig. The driller knows the value of teamwork and sees that the crew operates as a well-trained team; knows the ability of each individual; and won't assign a green hand a job beyond his or her experience.

The driller and the rig manager decide when to pull and change the bit. During active times on the rig floor, such as during a trip or a connection, the driller runs the rig at a pace suited to the crew, knowing that pushing them beyond their ability is dangerous. The driller makes certain the new crewmember knows the job, as well as the safety procedures involved in doing it.

The driller sees that the rig is well-maintained and trains the crew to be on constant alert for worn or faulty equipment. Once a problem is spotted, immediate steps are taken to correct it. Drillers make certain all guards, signs, and protective devices are in place and in good condition. They do not allow makeshift or shortcut operations that could be potentially dangerous. A driller must enforce the use of personal safety harnesses by anyone climbing or working overhead. Horseplay is not allowed on the rig. Should a crewmember show signs of being under the influence of alcohol or drugs, the driller will relieve him or her of duty immediately and report the occurrence to the rig manager. The driller keeps all important paperwork current, such as bit performance records, daily reports, and time sheets.

The crew usually consists of a derrickhand, two or three rotary helpers (also called roughnecks or simply "hands"), and, depending on the rig, a motorhand and electrician. Each must know the job and function as a team in order to work efficiently and safely (fig. 7).

Each member of the crew must report to work on time and be alcohol and drug-free. The job demands for everyone to be fully alert. Many contractors conduct random drug testing with mandatory testing of the whole crew following an accident. Crewmembers must wear personal protective equipment, including proper clothing. Long hair can be hazardous. In some situations, beards are forbidden.

On the rig, each crewmember must be safety-conscious at all times. The careful worker learns to be observant, foreseeing danger before harm is done. Some jobs are repetitive, which can lead to carelessness; that is when the danger increases.

As shown by figure 2, every crewmember should be aware of being struck by or caught between objects, especially around the pipe racks, mud pumps and tanks, and on the rig floor. A rig worker should not engage in unsafe practices. Tools must not be lying around the floor, ladders, or walkways because they present a trip or fall hazard. Crewmembers should always use the ladders and stairs to avoid jumping off anything over three feet high. When safety belts or harnesses are needed, wearing them should become automatic. No one should slide down rails or ride the catline or elevators. Very importantly, everyone helps the new crewmember by instructing him or her on the correct way to do the job safely and by pointing out the hazards around the rig. A competent crew avoids horseplay and practical jokes that might have serious consequences.

Crewmembers

Crewmembers must be alert and drug-free on the job.

Figure 7. Crewmembers working safely as a team

To summarize—

Responsibilities for safety on a drilling crew

- The operator selects a contractor with a suitable rig and may specify the casing and mud programs to be used.

- The drilling contractor devises a drilling program that follows government regulations and requires training for employees.

- The safety engineer, if there is one, investigates accidents and makes recommendations to the contractor.

- The drilling superintendent consults with the operator on a rig capable of handling the job, works with rig managers to make sure that equipment is rigged up safely, and conducts safety training for crews.

- The rig manager makes sure that operations are within the capability of the rig and ensures that all crewmembers are adequately trained. He or she also investigates accidents and eliminates hazards.

- The driller reports to the rig manager. He or she operates important controls and directly supervises the crew, setting a safe pace for work. The driller checks for safety equipment, as well.

- The derrickhand, rotary helpers, and other members of the drilling crew follow the safety practices for their positions, which include wearing personal protective equipment and being alert.

Personal Safety Equipment

In this chapter:

- Standard personal protective equipment for all crewmembers
- Specific requirements for safety-approved PPE items
- Personal items that cannot be worn on a rig
- Proper clothing for a rig
- Specialized equipment for certain operations

No matter how well-designed the rig or how well-supervised the crew, only careful, conscientious conduct by everyone can assure safe operations. Each person should receive instructions on the proper way to perform his or her work and the use of personal protective equipment (PPE). Safety standards prohibit the wearing of rings or other jewelry and loose clothing; they also prohibit long hair on a drilling rig. All these things are subject to being caught in moving machinery.

The mandatory personal protection items worn by each crewmember include a hard hat, safety boots, gloves, and safety glasses. Proper clothing is also important.

Hard Hat

PPE for all crewmembers:
- Hard hat
- Safety boots
- Protective gloves
- Eye protection

Safety hard hats must be worn by all personnel, as well as visitors in the work area. Hats must be of nonmetallic, nonconductive material (fig. 8). They must meet prescribed standards of strength and protection from impact, flying objects, or electrical shock. A bright color increases visibility.

Figure 8. Hard hats must be of nonmetallic, nonconductive material.

Safety Shoes and Boots

Steel-toed footwear made of heavy leather or neoprene construction is mandatory on drilling rigs (fig. 9). It greatly decreases the number and severity of toe injuries. Soles and heels should be of nonskid material, as rubber is too slippery. They should be discarded when the metal begins to show. Many contractors arrange a discount for their crews to purchase boots.

Figure 9. Footwear recommended for well site

12

Gloves protect the hands from minor injuries and improve the hold on slick or wet surfaces. They should be of heavy cotton construction and be short and close fitting. Large gauntlets are apt to get caught in something, so gloves of this type should not be worn. Periodically, the contractor may supply free gloves. Rubber gloves are needed when handling caustic or corrosive material (fig. 10).

Gloves

Figure 10. Rubber gloves are used for chemical handling.

Eye protection is of utmost importance. If an eye injury occurs, prompt medical attention is critical. Personnel should always wear approved safety glasses, goggles, or face shields when engaged in any activity where flying material is generated (fig. 11). These should be readily available on every rig. The risks to eyes are numerous, whether the worker is engaged in routine maintenance chores or a major activity like pulling the drill string. Any action involving a hammer can produce flying chips. Wire brushes, grinding wheels, scraping, compressed air, and steam cleaning operations all present dangers to vision.

Safety Glasses, Goggles, and Face Shields

Figure 11. Goggles or safety glasses are critical for eye protection.

Approved splash-proof goggles or face shields (fig. 12) should be worn when handling hazardous materials, such as caustic, cement, cleaning solutions, molten metal, or chemicals of any kind.

Electric-arc welding operations should be shielded to prevent damage to the eyes of observers. The welder should always have a helmet on, and helpers should be furnished with special goggles.

Contact lenses are discouraged because they can hinder attention to the eye if immediate action is needed.

Figure 12.   Face shield

Clothing

Only close-fitting, clean clothing should be worn. Long-sleeve shirts with shirttails tucked in are recommended. Cuffless trousers help prevent tripping. Each crewmember should have available a fresh change of clothing should his or her work clothes become oily or soaked with chemicals. Aprons should be worn when handling chemicals.

In cold weather, dress as lightly as possible considering the weather and activity level. It is better to be a little cold than too warm. Overheating can cause sweating, dehydration, and chilling when activity stops. Keep clothing as dry as possible and wear the clothing in layers. Sweaty or wet socks can lead to frostbite or trench foot; change them often.

Hearing protection gear should be worn around any high-noise area, such as around the rig engines (fig. 13).

Respirators are required for spray painting, mud mixing, or working in heavy dust (fig. 14). Special breathing equipment and instruction are needed if hydrogen sulfide is or may be present. This hazard is discussed in detail in later chapters.

Falls cause a significant number of injuries; therefore, a full-body safety harness is required for anyone working or climbing six feet or more above the derrick floor.

Figure 14. Respirator

Specialized Equipment

Figure 13. Hearing protection comes in all forms, but ear muffs are recommended.

Specialized PPE:
- Hearing protection
- Respirators
- Harnesses

To summarize—

Prohibited personal items on a rig

- Jewelry, including rings
- Overly loose clothing
- Long hair

Personal protective equipment for all crewmembers

- A hard hat of a nonconductive material that meets prescribed standards of strength and protection from impact and electrical shock
- Safety shoes or boots of heavy construction with steel toes and nonskid material
- Gloves of heavy cotton construction that are short and close fitting for most jobs; rubber gloves for handling caustic or corrosive materials
- Safety-approved glasses, goggles, or a face shield, depending on the task

Specialized personal protective equipment

- Hearing protection—for high-noise areas
- Respirators—for spray painting, mud mixing, or working in heavy dust
- Full-body safety harnesses—for climbing or working aloft

Safe Posture

In this chapter:

- The importance of good posture for safety
- Keeping the floor clean and dry
- The safe foot stance and posture for lifting
- Getting help from other crewmembers for heavy loads
- Using mechanical help for lifting

Strains, overexertion, and back injuries are a leading cause of *lost-time incidents* (*LTIs*), or accidents that take a person off duty. When lifting, good footing is of first importance. Clean up a slick or cluttered surface before lifting; otherwise, a slip or trip in the middle of a lift can be injurious. Good balance and posture are critical. Most strains and sprains occur as a result of an awkward, off-balance stance or by overreaching. A correct lifting posture is required. Squarely face the object to be lifted with feet spread a shoulder width apart. Bend the knees and test the weight and grip; then, with the back straight, lift with the legs while holding the load close to the body (fig. 15). Do not twist the back while moving the load, and set it down the same way it was lifted. Weight lifting competitions are not allowed.

Figure 15. Recommended way in which to lift an object

Whenever possible, crewmembers should make use of heavy lifting equipment to avoid strain injuries.

Never attempt to lift an object that exceeds your lifting ability (fig. 16). Always get help when moving a heavy or awkward load. It is better to have more than enough lifting assistance than to have too little. Coordinate the lift and set-down moves.

The best way to avoid back injuries is to use mechanical help for heavy lifting. That is what the catline, winch or boom, jacks, block and tackle, pinch bars, and hoists are for. Don't hesitate to use them.

Figure 16.   Personnel are urged to obtain assistance when it is required.

To summarize—

- When lifting, set the feet even with the shoulders, lift with the knees, and avoid twisting or overreaching.

- Do not lift an object that is too heavy or awkward; instead, get help from another crewmember.

- Use appropriate and safe mechanical help for lifting.

Offshore Transportation Safety

In 2010, a blowout occurred at the Macondo Prospect off the coast of Louisiana. In response to the many fatalities among workers on the rig and the leaking of millions of barrels of oil into the Gulf of Mexico, the U.S. federal government enacted new laws governing the industry. The Bureau of Safety and Environmental Enforcement (BSEE) is the agency that oversees worker safety and enforces environmental safeguards for offshore drilling.

Someone who has worked on a land rig will find many of the safe operating procedures for an offshore rig familiar. In the challenging offshore drilling environment, however, there are additional safety considerations. So the BSEE has set regulations for such matters as the transportation of crewmembers by boat or helicopter.

Begin your stay offshore by reporting to the dock or heliport on time (fig. 17). Check in early with the dispatcher to see that your name is on the passenger list. Provide all requested information: name, weight (including baggage), company, and destination. Advise the dispatcher and pilot if any type of authorized hazardous material is being transported. Do not loiter around the dock or heliport area. Remain in the designated passenger area until the pilot or captain has given clearance to board. Never attempt to take alcohol, drugs, firearms, explosives, or flammables onto an offshore installation.

Figure 17. Always report to heliport on time.

The pilot is in total command of the helicopter, passengers, and cargo. Follow instructions! A pilot can refuse passage to anyone thought to be under the influence of alcohol or drugs—or anyone who appears to be an unsafe passenger for any reason.

- Hold lightweight articles like hats, jackets, or raincoats firmly to prevent them from being sucked into the rotors. Carry long articles parallel to the deck, keeping them clear of the rotor blades.

- Approach and depart the aircraft from the front. Walk briskly but do not run. Avoid the area of the tail rotor and boom at all times. Approach or depart the helicopter in a crouched position, keeping well below the rotor tips (fig. 18). Be especially wary in high winds when the rotor blades may dip below six feet.

- Do not store luggage until the pilot gives you instructions.

- When boarding or departing, step carefully on the proper footholds to avoid damage to the craft's floats. Do not jump!

- The pilot is concerned with load distribution. Store baggage according to instructions and take the assigned seat. Fasten the seat belt and keep it fastened until told it is safe to remove it.

Helicopter Transportation

When approaching a helicopter, crouch down and be aware of the rotors and boom.

Figure 18. Proper crouched position for helicopter approach

- Inflatable life jackets should be worn during all overwater flights. Return the jacket to the storage area before departing the aircraft.

- Before takeoff, each person should note the closest emergency exits, inflatable life rafts, fire extinguishers, or other emergency equipment. Advise the pilot of any problems with the equipment at once. Pay attention to the pilot's preflight emergency briefing and follow instructions should an emergency occur.

- Smoking is not allowed in or around the aircraft at any time.

- Never throw anything from the helicopter. This could result in damage to the rotors.

- If hearing devices are provided, wear them as needed or instructed.

- In an emergency situation, remain in the seat with the seatbelt fastened. Do not attempt to jump from the craft; await the pilot's instructions. Do not inflate a life jacket or the inflatable life raft inside the cabin. Such actions would hinder or possibly prevent a speedy, safe evacuation.

Crew Boat Transportation

The captain is in command of the boat, crew, passengers, and cargo. Like the helicopter pilot, the boat captain can refuse passage to anyone thought to be under the influence of alcohol or drugs—or to anyone who appears to be an unsafe passenger for any reason.

- Smoking is not allowed in or around the crew boat.

- Passengers must ride inside the boat and not on the deck unless there is an emergency or they are specifically authorized to be there.

- The disembarking process is potentially hazardous, particularly in high winds or rough seas. Wait for the captain to give clearance to exit.

- There are several means of transferring passengers from the crew boat to the platform, including the personnel basket, transfer capsule, telescoping gangway, ladder or stairway, and swing rope. The most common is the personnel basket (fig. 19).

  Whichever method is used, follow safety instructions:
  - Securely fasten your life jacket.
  - Be careful with footing on wet surfaces.
  - Keep knees braced for unexpected movements.
  - Place luggage and other materials where designated.
  - Avoid getting feet or legs caught between the boat and the platform.
  - Avoid arranging ropes in a way that could entrap body parts.

- After coming onboard, report immediately to the person in charge of the facility and sign in. You will be assigned a bunk, locker, and duty station. Information about the rig will be provided: emergency signals, communications, first-aid equipment, reporting injuries, escape routes, smoking regulations, wearing work vests or life jackets, overwater transfers, and fire systems. Procedures for reporting leaking fuel or escaping oil or gas will be explained.

- Specific duties in the event of an emergency, fire, or abandonment will be explained. These emergency procedures are posted on station bills located throughout the facility. Read the bills carefully and be prepared to perform the procedures if necessary.

- Pollution prevention is especially important offshore. Throw nothing overboard. All rules designed to protect the environment should be strictly observed.

Figure 19. Basket-lift transfer to an offshore platform

To summarize—

- When traveling to an offshore location in a helicopter, follow the pilot's instructions. Follow recommendations for posture approaching the helicopter to avoid being struck. Hold personal articles close to the body to avoid being caught in the craft's moving parts.

- Do not bring flammable or explosive materials on board a helicopter or a boat. Do not smoke on board.

- Do not report for travel after drinking alcohol or taking intoxicants. Do not bring alcohol or drugs on board.

- When traveling in a crew boat, do not venture out onto the deck.

- When disembarking to board the rig, follow safety procedures appropriate for the transport unit.

Hand-Tool Safety

In this chapter:

- General safety rules for commonly used hand tools

- Special precautions for using individual hand tools

- Tips for storing tools and for transporting and working with them aloft in the derrick

- Precautions for using portable ladders

A member of a rig crew must be able to safely use a wide variety of tools. Many will be similar to home tools but more heavy duty. Keys to hand-tool safety include:

- Use the proper tool for the job.

- Inspect the tool before using it. Be sure the tool is not worn, broken, or damaged. Report damaged tools to the driller.

- Never use a tool a task it is not intended for. Do not use a wrench for the job of a hammer, for example.

- Be certain the area is clear of people and obstacles when swinging any tool.

- Maintain a good stance so the tool does not slip. Do not overreach!

- Carry tools safely. Tools slip from pockets, so use a tool belt, especially for sharp or pointed tools.

- If you don't know how to use a tool, ask for instructions.

Having the right tool readily available for a specific job reduces the temptation to use the wrong tool, which can cause an accident. Tool boards located around the rig provide an easy way to keep tools accessible and in good condition. Each crewmember should take the initiative, seeing that tools are not left lying around the rig, creating a hazard.

Time is saved and hazards avoided when tools are kept in their proper spot on the board. In that place, dirty, missing, or damaged tools are readily spotted (fig. 20). Drillers should check the tool boards frequently to see that any defective tool is replaced or repaired as soon as possible.

No one should be on the floor when someone in the derrick is working with hand tools. If a tool is dropped from the derrick, a serious accident may result. Tools in the derrick should be tied to a rig member or placed in a toolbox securely attached to the rig. All tools should be lowered to the derrick floor when no longer needed.

Figure 20.   A well-organized tool board

Hammers

Hammers are one of the most-used hand tools, and their proper use is important to safety. Hammers with damaged heads or handles should not be used until repaired. It is recommended that a crewmember using a hammer, as well as anyone nearby, wear safety goggles. Be sure everyone is at a safe distance from the hammer user. A missed or glancing hit may cause an injury to the user or someone nearby. The hammer should be swung with both hands held close to the end of the handle. Choking the handle results in increased danger to hands and fingers as well as a less effective hit. Handles should never be used as prying tools. Do not strike hardened objects (such as wrenches, files, and other hammers) with anything other than a rawhide or soft-metal hammer. An inexperienced or undersized worker should not use a sledge. A 140-pound floorhand should not normally be using a 16-pound sledge.

Wrenches

Wrenches may be of several types: open-end, box-end, adjustable (Crescent™), socket, pipe, or special design, like a hammer wrench or tong wrench. The key to the safe use of any wrench is in selecting the proper type and size for the job at hand. Check all tools before use for worn or sprung jaws, bent handles, missing springs or faces, and broken cages. Promptly advise the driller of any such damage.

Correct use of a wrench will avoid an accident (fig. 21). If location and space allow, it is always safer to pull on a wrench than to push on it. If the wrench slips while being pushed, a loss of balance is more likely. Heavy pressure is put on the jaws when a nut or fitting is tightened. Thus, it is important to see that adjustable wrenches are set with a full, snug fit on the object being tightened. A loose fit may cause slippage, hand injuries, scratches, or the rounding of a bolt or nut. On any wrench, the fixed jaw is stronger than the movable jaw, so direct the force toward the movable jaw. This tightens the grip and prevents the jaws from spreading. With end wrenches, the box end will give a more secure grip and should be used to break out a tight nut or to make the last turns when tightening.

On large pipe wrenches, make sure the jaws have a firm grip before applying weight. Use a tong wrench on pipe too large for a pipe wrench. Place your feet so that you will not lose your balance should the wrench slip. Stance is especially important in case you have to push, rather than pull, on the wrench. If more leverage is needed, use a larger wrench. Cheater bars may break a small wrench but they may be safely used on larger pipe wrenches. When extra leverage is needed, use of a cheater bar is preferable to standing on the wrench. Standing or jumping on pipe wrenches has caused many leg injuries.

Allen wrenches should never be used as a retainer on a pressure-relief valve because they are made of tempered steel and could cause excessive pressure buildup and explosion.

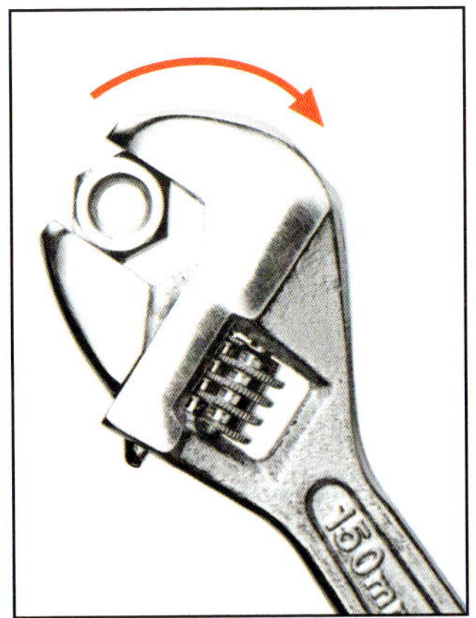

Figure 21. The pulling force on a Crescent™ wrench should be on the nonmoving side of the jaw.

Pliers

Pliers are usually of three types: slip-joint, vise grip, and adjustable. Generally, pliers are intended to be one-handed gripping tools rather than replacements for wrenches to tighten or loosen bolts and nuts. Because pliers have flexible jaws and their hold depends on the user's grip, they are more likely to slip. Injured hands and the rounding of nut or bolt heads may result. For efficiency, the proper size tool and the positioning of the slip joint is critical. To maintain the best hold, the jaws must be as nearly parallel as possible. The adjustable settings allow pliers to achieve a firm hold on objects of various sizes.

Screwdrivers

The size and condition of a screwdriver's blade is an important safety consideration (fig. 22). The sides of the blade should be straight and exactly parallel in order to fit the screw head properly. The end of the blade should be blunt and at exact right angles, not sharp or chipped. The handle should be straight and smooth, but not slippery.

Whether the screwdriver is a straight (standard) or cross-blade (Phillips) type, the closeness of fit between its blade and screw head is very important. An improper fit is the chief cause of screwdriver accidents. For safety, use the largest screwdriver that will fit snugly into the screw head. Do not place a free hand where it may be struck if the blade slips and use a vise rather than a hand to hold small objects. Use a screwdriver with an insulated handle for electrical work.

A screwdriver should not be hammered on or used as a chisel or pry bar. Do not use a wrench on a screwdriver unless it is the heavy-duty type with a square shank.

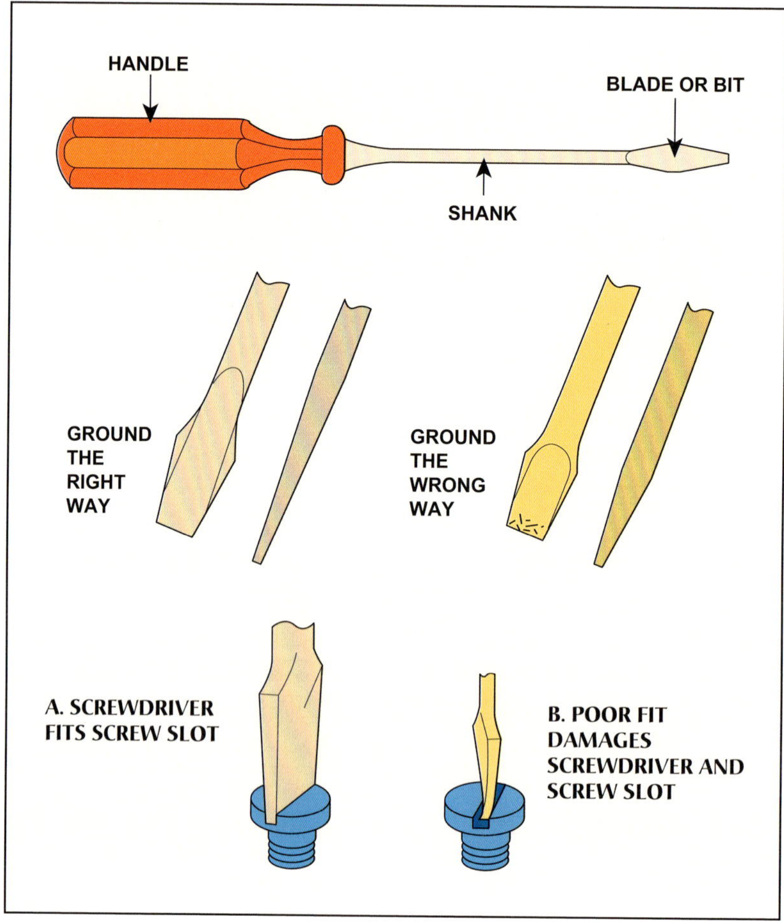

Figure 22. Screwdriver safety

Chisels present a danger to eyes and hands. Wear safety goggles or other approved eye protection when using this tool. Hold the chisel with the thumb and forefinger and with the palm up to avoid striking the knuckles (fig. 23).

For efficient, safe use, the chisel and hammer must be properly matched. A hammer must not be greatly oversized for the chisel because this lessens control of the tools. Always chip with the chisel pointed away from the body. Be sure the cutting edge is sharp and of the right shape. The head of a chisel may split, flatten, or mushroom with age. When struck, a tool in this condition can send off flying chips. To avoid that danger, the chisel head should be reshaped with a grinding tool prior to use.

Chisels

Make sure that tools are properly sized when they are to be used together.

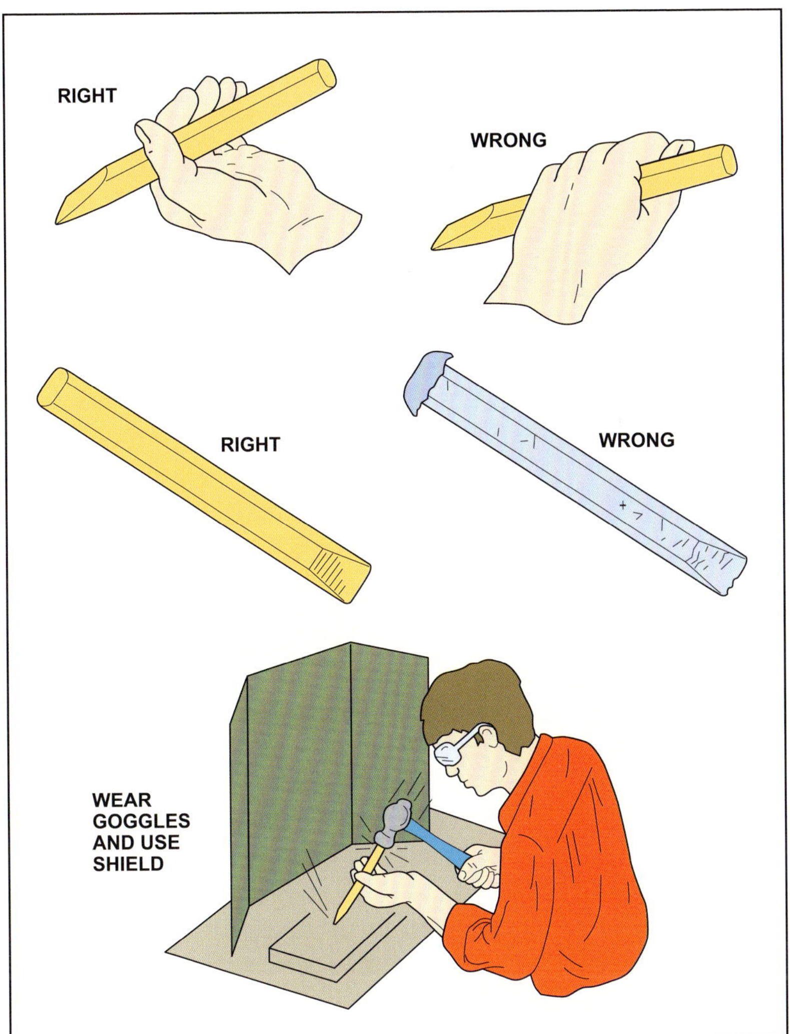

Figure 23.   Proper grip and maintenance of chisels

Use a vise rather than a hand to grip objects to be worked on with another tool.

Other safety considerations:

- Clamp small objects in a vise to use a chisel on them and chip toward the solid, or stationary, jaw of the vise.
- Never attempt to hold a chisel with hands that are numb from cold.
- Never use a wood chisel that has no handle.
- If another person holds the chisel, it should be held with a special set of pliers or other approved clamping device.

Files

Files are not to be used unless they are equipped with tight-fitting handles. The tang end of a file is often sharp, and a loose-fitting handle can slip off, resulting in injury. If possible, the object being filed should not be hand-held but rather held in a vise or in another secure device. The teeth of a file are made to cut in one direction only—forward and away from the user. Remember: A file is a hard, brittle tool. It may snap when bent, so it should not be used as a pry bar and it may shatter if hammered, sending chips in every direction.

Shovels

Shovels may cause back strain if improperly used. Slide one hand towards the scoop before lifting, then move the feet to avoid an awkward, twisting movement when emptying the shovel.

Brushes

Brushes should be used properly. Wear safety glasses when using a wire brush and always stroke away from the body. When using a dope brush, be aware of pipe movement. The brush should have a guard to prevent it from being dropped down the pipe.

Portable Ladders

Portable ladders should be inspected before use to make sure they have secure, nonskid feet. They should also be free of cracks, splits, and loose rungs. Never use a portable ladder in a horizontal position as a scaffold. Do not use aluminum ladders or ladders made of other conductive materials around electrical lines. Personnel should not climb higher than the third rung from the top on extension ladders or the second rung from the top on folding ladders. Only one person at a time is allowed on a ladder. Wooden ladders should not be painted because the paint may cover serious flaws.

To summarize—

General rules for the safe use of hand tools

- Use the right-size tool that is intended for the job.

- Check the tool for damage and wear and replace or repair it when necessary.

- Make sure the work area is clear before using the tool.

- Do not place your hands in the way of the tool.

- Carry the tool safely.

Special rules for the safe use of common hand tools

- When using a hammer, wear safety goggles. Do not use a sledge that is too heavy or strike hard objects (including other tools) with a hard hammer.

- When using a wrench, select the right type for the job. Set an adjustable wrench tightly and pull toward the movable jaw rather than pushing toward the fixed jaw.

- When using pliers, set the jaws as close to parallel as possible. Do not use pliers in the place of a wrench.

- When using a screwdriver, select the one with the largest blade that will fit into the screw head. Use a vice (rather than a hand) to hold small objects, and do not use a screwdriver as a chisel or a pry bar.

- When using a chisel, wear safety goggles and position the hand correctly to grip it. Match the hammer to be used with it for size.

- When using a portable ladder, check it for damage. Do not use it in the horizontal position or climb too high on it. Do not use ladders made of conductive materials around electricity.

Power-Tool Safety

Portable power tools are often used around a rig. They may be pneumatic (activated by compressed air) or electrically powered. Typical injuries from power tools include cuts, burns, and electrical shocks. Eye injuries and falls also occur when using power tools. Power tools can even cause gas explosions. Falling tools can also cause injury. Many injuries can be prevented with simple precautions:

- Electrical lines on the ground or rig floor may cause someone to trip. Flag them or string them overhead.

- Do not leave power tools aloft. Moving equipment or personnel on any level may pull the power line, causing the tool to fall.

- Keep all power lines away from hot surfaces like mufflers or ones undergoing welding operations. Lines should also be kept away from gasoline, oil, and chemicals.

- Inspect all lines—air or electrical—before use. Check closely for loose connections. Any worn, frayed, or kinked areas should be promptly repaired.

- Do not disconnect a power line while a tool is in use. Loss of power can jam the tool, exposing the user to injury. Likewise, never plug in to an electrical socket or turn on the air without determining that the person using the tool is ready.

Air Tools

Air tools operate on about 90 pounds per square inch (psi) of pressure. A disconnected air hose will whip about, possibly causing damage or injury. An air hose should not be disconnected from a tool. To clean equipment, a cleaning air hose should come from a separate tank or compressor at about 30 psi. If using an air hose to clean machinery, safety glasses are needed. Never use an air hose to clean clothing.

An air gun or jackhammer has the cutting or riveting tool fitted into it. A rapidly moving piston impacts on the tool, which in turn impacts on the objective. Two safety devices are critical:

- The trigger should be located inside the handle, where it is safe from accidental activation. It should never operate until the trigger is depressed.

- A device should be in place that prevents the impact tool from being shot from the barrel. Small air guns may not have these safeguards. Therefore, a good safety rule for air hammers is "Don't squeeze the trigger until the tool is on the work."

Air and impact wrenches create high torque, so a firm grip must be maintained (fig. 24). These tools can be injurious unless carefully controlled. Hand-tool sockets should never be used on impact wrenches.

Air drills, grinders, and scaling tools have built-in safeguards to protect the operator, but the operator should be instructed in their use. These tools create flying material, so safety goggles should be worn when using them.

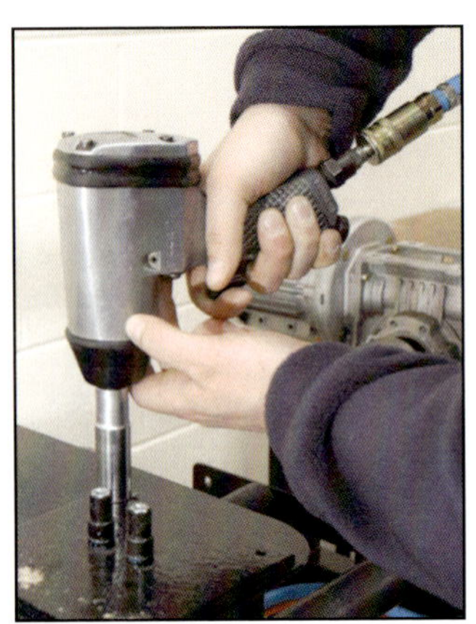

Figure 24. Maintain a firm grip on air and impact wrenches.

Proper grounding of electrical tools is critical. If there is a short circuit inside the tool, the current is drained from the metal frame through the ground wire.

The best way to ground an electrical tool is to use a three-wire plug and receptacle. The third wire terminates in a built-in ground, so the operator need not worry about attaching the ground wire. Never stand in water when using an electric power tool, even if the tool is grounded or double insulated. Pull on the plug, not the cord, when unplugging a power line. Heavy-duty plugs clamped to the cords help prevent strain on the electrical elements should workers wrongly pull on the cord rather than the plug.

Electric drills and saws should all be equipped with automatic cut-off when the trigger pressure is released. If an electric drill is dropped or a drill bit breaks, the eyes may be endangered. A face shield or goggles provides protection from such events.

Electric grinders, buffers, saws, and brushes should be shielded as much as possible, and the user should wear eye protection (fig. 25).

Electric Tools

Avoid electric shock. Ground all power tools and do not use them while standing in water.

Figure 25. Always wear eye protection when using electric grinders.

Do not force electric saws to cut when they jam, and never remove mechanical guards. If excessive dust is created, a breathing mask is advised. Promptly clean up any debris created by the operation.

Electrical tools should not be used in a gaseous atmosphere or near flammables because they create a danger of explosion or fire.

Gloves should not be used with some hand tools that generate a rotary motion because they can become entangled in the rotating parts.

To summarize—

Types of power tools

- Pneumatic, or actuated by compressed air

- Electrically powered

General rules for the safe use of power tools

- String power lines overhead or flag them for the crew. Keep them away from hot surfaces and flammable materials.

- Be careful using power tools aloft in the derrick and while lifting and lowering them.

- Inspect tools and power lines before use and make repairs, if necessary.

- Do not disconnect the power source while a tool is being used or power up a tool before the user is ready.

- Wear goggles while operating tools.

Special rules for the safe use of common power tools

- When using air tools, do not disconnect the air hose while the tool is operating. Grip the tool firmly and follow precautions for using the individual tool to avoid activating it at the wrong moment.

- When using electric tools, make sure that they are properly grounded and shielded. They should also have an emergency shut-off. Do not use electrical tools while standing in water or in an area where gas is in the air.

Rig-Up Safety

In this chapter:

- Dangers of rigging up
- Why rigging up is only allowed during the day
- The importance of being alert and staying out of the way
- Special precautions for raising derricks, positioning equipment, and testing engines

Lost-time injuries (LTIs) occur most often when activity on the rig is intense. The most dangerous activities are equipment repair and maintenance, routine drilling operations, tripping in and out, and rigging up and down. During rig-up, there are many hazards (fig. 26):

- Getting caught in a *pinch point*
- Getting struck by a crane, truck, load, or falling tool
- Falling from a height
- Becoming entangled in lines
- Getting crushed by equipment being put into place

Figure 26. A forklift can crush a person or tip over.

- Slipping on wet surfaces or debris in a walkway
- Getting shocked or electrocuted

Rigging Up

Rigging up begins after the operator has prepared the location with pits, roads, and a level surface designed to accommodate the specific rig to be used. Site hazards, such as pipelines, overhead electrical lines, or airport approaches, must be dealt with. The drilling contractor's superintendent or rig manager contracts out the work of trucking in the rig, in addition to supervising the installation of the cellar and conductor pipe and the spotting and layout of the rig. A welder, crane, and other service equipment will be available as needed.

The actual rigging up of the typical land rig involves two or three crews working together at the same time under the direct supervision of the rig manager. They begin as soon as the equipment arrives at the location, and rigging up may take several days for the larger rigs. It is done only during daylight because of the dangers of falling and getting caught between or struck by objects. The derrick is assembled in a lay-down position, drawworks and blocks are strung, and motors are activated. Using power from the drawworks, the mast is then raised (fig. 27) and secured. At that time, crews begin work on a 24-hour, two- or three-tour schedule.

Figure 27.   Raising a drilling rig mast

All tools and loose equipment of any kind must be removed from the derrick before it is raised to the vertical position. Absolutely no one should be on or under the derrick while it is being raised because the possibility of equipment failure is always present. After the mast is raised to its vertical position and all pins are in place in the substructure, workers can begin rigging up. Before moving any equipment to the floor, secure all floorboards in place and cover all open holes. All handrails, walkways, ramps, and stairs should be bolted or pinned in place as soon as the rig equipment is positioned on the derrick floor.

Everyone must be especially alert in order to avoid placing themselves or others in danger during the rig-up process. Rigging up is hazardous because many people are working at different nonroutine jobs simultaneously—sometimes in close quarters—while heavy machinery and equipment is being moved about. Some of the crew may be working on the mast while others are installing mud pumps or mud tanks, hooking up pipelines, placing the drawworks, adjusting engines, and installing blowout preventers, shale shakers, and guy lines, or performing a dozen other tasks. There is added danger from equipment being hoisted overhead. Some crewmembers may be inexperienced, or not familiar with a particular rig or item of equipment. All these factors increase the risk of accidents.

If engines must be started or tested before all belt and chain guards are in place, special precautions must be taken to see that all personnel are clear of rotating equipment, mud pumps, and the drawworks.

Never walk or stand under anything being lifted overhead. Use taglines to guide a load. When loads are being hoisted, there is always the possibility that a line may slip or break, objects may fall from above, or broken lines may whip about dangerously. Be prepared for the sudden movement of anything being handled by winch line, hoist, crane, or jacks. Keep feet and hands clear of potential pinch points, where they could be crushed by an unexpected shift or by dropped equipment.

Be careful with jacks! Be sure the base of the jack is placed on something that will safely bear the load. Do not leave the handle in a jack that is supporting a load.

Because serious injuries are more likely to occur during rig-up, that operation is only conducted during daylight.

Never hammer or try to disconnect anything that may be under pressure. Be certain all pressure has been bled from air hoses, pipelines, hydraulic lines, vessels, regulators, or connections before working on them. When removing pipeline supports, hold the line by hand, if it is lightweight, or by hoist or winch line if it is heavier. Pipe fittings and connections must be made up tightly but carefully if using power tools. These tools can quickly overtighten, twisting off nuts, bolts, or connections.

Observe all the general rules previously discussed concerning safe posture, personal protection equipment, and hand- or power-tool safety.

To summarize—

Dangers during rig-up

- Rigging up is one of the most hazardous operations conducted on a drilling rig.

- Tasks during rig-up are diverse, and so are the injuries sustained while doing them. They include getting caught in a pinch point, being struck by moving equipment, falling, and being electrically shocked.

Rules for a safe rig-up

- Manage hazards at the new site such as overhead electrical lines.

- Only perform the rig-up during daylight hours.

- Before raising the derrick, remove all loose objects from it.

- After raising the derrick, repair uneven flooring. Install walkways, ladders, and handrails.

- Before moving large objects or starting the engines, make sure people are cleared.

- Be alert. Stay clear and keep away from pinch points.

Drilling Operations and Equipment Safety

▼
▼
▼

In this chapter:

- Keeping drilling equipment in position
- Preventing falls of equipment and people
- Keeping things from getting entrained by equipment
- Special precautions for using drilling equipment
- Maintaining drilling line and slips to avoid accidents
- Preventing strain by using power equipment

Depending on the size of the rig, the crew usually numbers from four to six people. These people are the driller, derrickhand, motorhand, and two or three floorhands. The derickhand is under the direct superision of the driller; the entire crew is under the overall supervision of the rig manager (toolpusher). Offshore crews may be supplemented by a crane operator, roustabouts, a mechanic, and an electrician. Floating drill vessels employ a subsea engineer, barge master, and watch standers.

The basic crew is responsible for normal drilling operations, maintenance, and repairs of the rig. It may also run casing, handle blowout prevention, and do completion work. The crew is often assisted by contributions from many oilfield specialists: mud engineers, fishing tool operators, directional drilling operators, pressure control engineers, casing crews and cementers, and logging and perforating personnel. The operator may also employ an on-site consultant.

After drilling has reached total depth and downhole logging has been completed, a critical decision must be made. If the hole is cased and completed, will it be a commerical well that justifies the cost?

If the decision is no, then the well is plugged and abandoned. If it is yes, then casing is run and, in most cases, the drilling rig is released after the production casing has been set unless conditions require the drilling rig for completion operations.

Derrick or Mast

Figure 28. Installing guy line anchors

Secure the derrick or mast with lines and tie back the stands of pipe in the fingerboard to prevent those components from moving in the wind.

All derricks or masts must be protected from being overturned by high winds. The danger is increased when pipe is racked in the mast. For masts that require guy wires, the industry has developed generally accepted safe guying procedures (fig. 28). A guying system constructed and installed according to the manufacturer's specifications virtually eliminates the danger of an overturned mast. The guy lines must be properly anchored and clamped prior to the start of drilling operations. Flagging the lines near the ground and at head level helps the crew avoid running into them in poor light.

For portable telescoping masts, these general rules apply:

- A telescoping mast must be equipped with a safety device that engages automatically should the lifting mechanism fail. The device will prevent the upper section from collapsing while being raised or lowered.

- Install external guy lines except when handling very light loads. Attach guy wires to permanent, buried ground anchors (deadmen) if possible; if not, use expanding or screw-in anchors.

- All guy lines must have enough clamps and be properly fitted and tightly fastened to prevent slippage.

- If a pull lift (come-along) is used to tighten a guy line, tie down the handle after the final pull.

- If raising or lowering operations do not proceed by regular hydraulic or drawworks power, do not apply additional force with winch lines or other means.

- Be sure the lock nuts on all jacks supporting the structure are in place before the mast is raised. They must remain in place until the mast has been lowered back onto its cradle.

- To level a mast, all jacks must have a solid footing. Set them on a support beam, timber, or mat.

- Line up the mast by using the jacks, not the guy lines.

- If a rig must be left standing with pipe in the mast, put a sling around the wellhead and latch or chain the brake drum to it.

- Inspect all jacks and jack locknuts, pull lifts, guy lines, and clamps regularly for tightness.

The mast's pipe-racking support is designed to support the top of the pipe stands. It should be stoutly constructed and, along with the mast, should completely enclose the pipe. The pipe fingers should be straight and secured with a safety device to prevent them from falling if they break. On each trip, the derrickhand (fig. 29) should inspect the monkeyboard, tieback ropes, and escape device (Geronimo) as well as the safety lines attached to all sheaves hanging in the derrick (including the tongline and catline).

Stands of pipe should be tied back in the fingerboard as soon as they are racked to prevent the derrickhand from being knocked off the monkeyboard. Pipe that is securely tied back will not shift in the wind. Pipe shifting in the mast puts great strain on it. No one should be in the derrick when pulling on or jarring stuck pipe.

Every derrick or mast should be equipped with sturdy ladders leading from the floor to the crown block platform. Ladders should be securely fastened in place (fig. 30) and constructed with uniform, parallel-rung spacing. A ladder should never lean backward. Face the ladder when climbing up or down.

Figure 29. Derrickhand on monkeyboard

Figure 30. Poster showing what could happen if rig stairways are not properly attached to a stable structure

Secure tools before raising them in the derrick.

Platforms are placed inside the derrick at each elevation where a worker will be handling pipe or other equipment stacked in the derrick. The inside edge of the platforms should allow safe passage of the traveling block but allow the person on the platform safe access to the elevators. Inside platforms except monkeyboards should completely cover the space back to the derrick girts. All platforms and ladders should be fastened to the derrick with bolts or other secure devices to prevent shifting or dislodging. Inspect the fastenings frequently. Alert the driller if any platform or walkway 10 feet (3 metres) or higher above the rig floor has a missing or damaged toe board. Check the fastenings frequently.

Derrickhands must use a climbing device with counterweight to get to their station (fig. 31). The climbing device has a friction control that prevents a rapid descent in the event of a fall. Personnel must never ride the elevators or be lifted by catline to stations aloft. Keep both hands free when climbing up or down any derrick or mast ladder.

Figure 31. Counterbalanced climbing device attached to a cable on the side of the rig

Any tools to be used overhead must be raised and lowered in a canvas bucket by catline, hoist, or handle, or they must be secured directly to the elevators or hook. Personnel on the floor must be alerted that tools are being used overhead. Precautions to prevent accidental dropping should be taken by anyone using tools in the derrick. Tools should not be left in the derrick when not being used.

All rigs must be equipped with a derrickhand's escape line, generally called the Geronimo or tinkerbell line (fig. 32). This specially rigged line and braking device gives the derrickhand a means of escape should a blowout, fire, or other emergency cut off escape by the mast's ladder and it must be in place before the first trip. The escape line should be installed according to the manufacturer's specifications. It should ideally be stretched at about a 30-degree angle to the ground; steeper angles give a too-rapid descent. The line must be deeply anchored to support the load created by a suspended worker. Escape lines are for emergencies only. The braking device is designed for only two or three descents before excessive wear affects their performance; therefore, they should be inspected before each trip.

A safety belt with harness should be provided to each employee working more than 6 feet (1.8 metres) above the rig floor (fig. 33). The belt and harness should be adjusted to fit snugly. The lifeline should be of ⅝-inch nylon rope or equivalent attached to the employee's belt and a derrick member. The recommended working length is 5 feet (1.5 metres), allowing enough movement to do the job, but no excess slack. Tie-off lanyards, or lifelines, should be equipped with a shock absorbing fall-arrest system.

Figure 32. Derrickhand attaching the Geronimo to a cable. The brake handle should always be on the right as the derrickhand faces outward from the back of the rod basket or tubing board.

Figure 33. Safety harnesses are required for anyone working more than 6 feet above the rig floor.

Special equipment that
protects the derrickhand:
· Escape line
· Safety belt and harness
· Lifeline
· Tail rope

The derrickhand's tail rope (fig. 34) should not be attached to the same girt as the collar and drill pipe snubbing lines. If it is necessary to move to the opposite side of the derrick, reattach the lifeline at another place before stepping off the monkeyboard.

Bending of any derrick or mast member can weaken the structure; therefore, chains, hoists, or snatch blocks should not be attached to rig girts. Tong safety lines should not be attached to mast or derrick legs unless the mast or derrick is designed to support them.

All stairs and guardrails on or around the rig must be maintained in good condition, secured in place, and properly lighted. If they are moved temporarily to allow installation of equipment, they must be replaced as soon as is practical.

Figure 34. Working from a platform called the stabbing board, the derrickhand, with a safety harness, escape line, and tail rope, guides the casing elevators near the top of the casing joing.

Only designated personnel should operate the drawworks or rotary table (fig. 35). However, all rig personnel should be instructed in how to shut them down in the event of an emergency.

The throttle for the engines driving the drawworks and rotary table must be designed and installed to provide the driller safe, effective control of the power source. The steel guard in front of the drawworks drum prevents anyone from falling onto the drum or being caught by a hoisting line. Do not stand on top of the drawworks or inside the drum or sandline guards while the drums are moving. There should be no exposed keyways or keys that can catch clothing on the drawworks or rotary table. Place guards over such hazards. All gears and chain drives must have strong steel guards securely bolted or screwed into place whenever the machinery is in motion. Do not distract the driller while the drawworks is engaged. The driller supervises all drawworks repairs, and no repairs should be started until the equipment is stopped and controls are locked and tagged. Brake components should be inspected frequently to insure against excessive wear. All parts of the brake controls should be guarded to prevent contact with a moving chain, belt, or line that would interfere with brake operation. Operators should chain down the brake control anytime they leave the driller's console. Do not direct water at the brake band when washing the drawworks.

Drawworks, Rotary Table, Rotary Hose, and Kelly

Drawworks

Figure 35. Drawworks

Rotary Table

Figure 36.   Rotary table in motion

Many accidents happen on the rig floor on or around the rotary table. The rotary table should not be engaged until all floorhands are clear. Never step on the rotary table while it is rotating (fig. 36). When spinning pipe out during a trip with the rotary table, be aware that downhole torque can cause the rotary table to rotate backward. Slips and falls are common on the rig floor. Keep the rotary table clear of all tools, hoses, lines, or ropes. The rotary table should be surrounded by a working surface constructed of or covered with a nonskid material (fig. 37). The area where the floorhands work should be level with the pipe racking area. This makes handling pipe easier and safer. Keep the mousehole and the rathole covered unless it is occupied by the kelly or pipe. When removing the rotary table, be sure the lift lines are strong enough. Space the lines evenly around the hood to assure a level lift. Keep pipe tongs securely tied away from the rotary, swivel, and kelly.

Figure 37.   Rig floor

The standpipe end of the rotary hose should have a clamp fastened to the mast or derrick with a substantial chain or cable. The swivel end of the hose should also have a clamp fastened to the swivel by a chain or cable (fig. 38). The chain or cable should not be attached to the gooseneck. The ends of connecting hose sections should be secured by clamps and a safety chain or cable in case there is an accidental disconnect. If a steel drilling hose is used, each section should be clamped and secured to the derrick to prevent its whipping about in case of a break. The safety lines should be anchored to the derrick near the swivel.

Rotary Hose

The rotary must be clear of people and tools when it is operating.

Figure 38. Rotary hose with cables and clamps

49

Figure 39. An air-powered kelly spinner

Kelly

A platform should be provided for a worker to stand on when the kelly is racked in the mast or derrick. A kelly slide, if used, must be well constructed and braced so that it cannot be displaced. The kelly valves must have a pressure rating that exceeds any pressure expected during drilling or completion operations. The key or handle for the kelly valves must be conveniently located on the derrick floor.

An air- or hydraulic-powered kelly spinner is safer than a spinning chain (fig. 39). It produces a more positive action when making a connection and there is no physical labor required of the floorhands. Mud flies about during a well kick. The kelly spinner can quickly make up a connection under these conditions despite the hazards from flying mud.

Accidents happen when machinery is operated at an excessive speed. The driller controls the pace of operations by controlling the drawworks (fig. 40). A competent driller maintains a safe, methodical speed. The drawworks should be slowed as a stand of pipe is being picked up so the derrickhand can assure that the elevator latch is safely fastened and the floorhands can attach a snubbing line. If the drawworks is not slowed, a stand of pipe will be picked up with a jerk, causing it to whip about and make control of the pipe more difficult, which presents obvious and serious danger to the floor workers.

Maintenance or lubrication should not be performed on the drawworks, swivel, or rotary while they are moving.

Figure 40. Driller at the controls of the drawworks

For safety, the crown block on a derrick should be raised and installed only during daylight hours. On portable rigs, the drilling line should be threaded through the crown and traveling block sheaves while the traveling block is on the rig floor supported in a vertical position or while it is lying on the catwalk. The drilling line should be installed so that the deadline is near the working platform and the fastline is away from the platform. Carefully inspect the crown (fig. 41) and traveling blocks before the mast is raised.

All rigs are equipped with safeguards to prevent the traveling block from hitting the crown block. The most common is an air-operated anti-crowning device commonly known by the brand name, Crown-O-Matic®. When a preset amount of drill line is spooled onto the drawworks drum a lever on the drum shuts down the power to the clutches and simultaneously locks the brake. The lever must be reset and air pressure bled off before the traveling block can be moved. Bumper blocks installed at the crown provide a second safeguard. These blocks prevent the traveling block from being raised too high and damaging the sheaves or drilling line.

Report any vibration or movement of the crown block assembly, then promptly tighten all bolts securely. All bearing caps must be tightly bolted to prevent the sheaves from jumping out of the block housing and falling.

Crown Block, Traveling Block, Hook, and Hoisting Line

Crown Block

Figure 41. Crown block showing sheaves

Traveling Block

Every traveling block must be equipped with heavy metal sheave guards. The guards are designed to prevent an employee's hand from being drawn into the area where the line makes contact with the sheave (nip point). Sheave guards must be securely fastened to the block. There should be no protruding parts on the block that could catch on clothing or other equipment. The drawworks must be closed down and the brake chained down when the crown or traveling blocks are being greased or repaired.

Hook

Check the safety latch on the drilling hook (fig. 42). The safety latch should be of a well-constructed positive action design to prevent it from disengaging under load. The latches should be fully closed and locked during trips so that a jar from the elevator links cannot drive the latch aside and unhook the elevator bails. A wireline sling should be used on the elevator bails if the hook lacks positive latches. The kelly should remain in the rathole until the swivel bail has been locked into position on the hook. Repairs to the swivel, gooseneck, or rotary hose (fig. 43) should be made only while the kelly is in the rathole, not hanging from the hook.

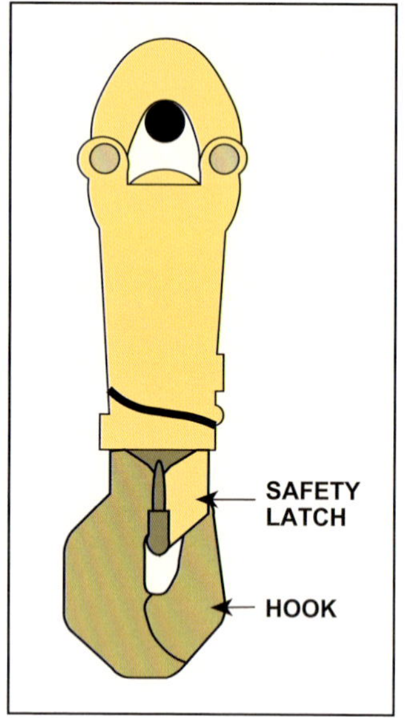

Figure 42. Drilling hook

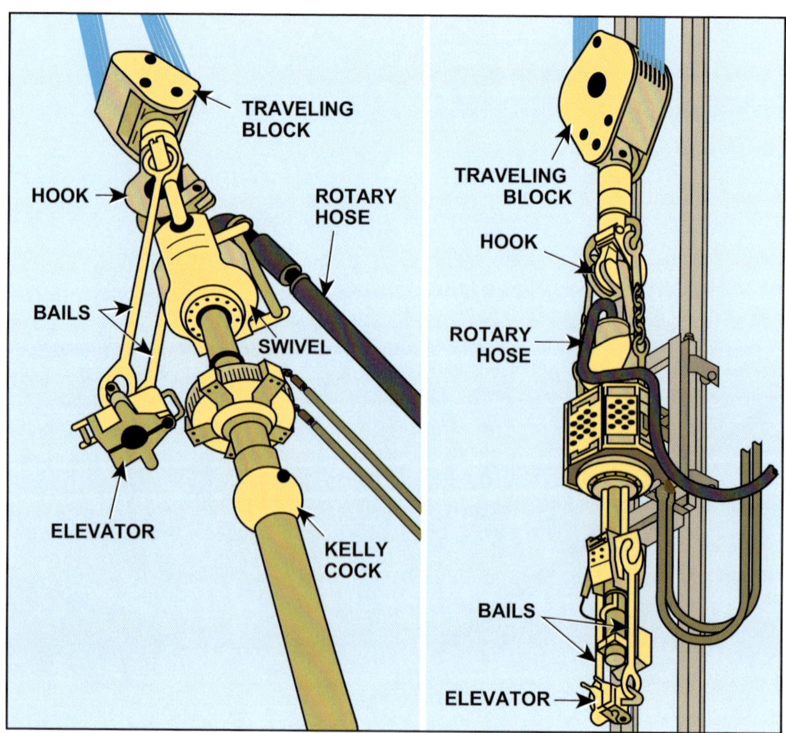

Figure 43. Traveling blocks and swivel assemblies

The dead end of hoisting lines must be secured to the rig substructure by a tie-down anchor, or deadline anchor (fig. 44). Two or more turns should be made on the anchor drum before clamping. All components of the deadline anchor must equal or exceed in strength the breaking strength of the hoist line. A bolt or retainer should be attached to the anchor to prevent the line from jumping the drum in case slack develops.

Hoisting Line

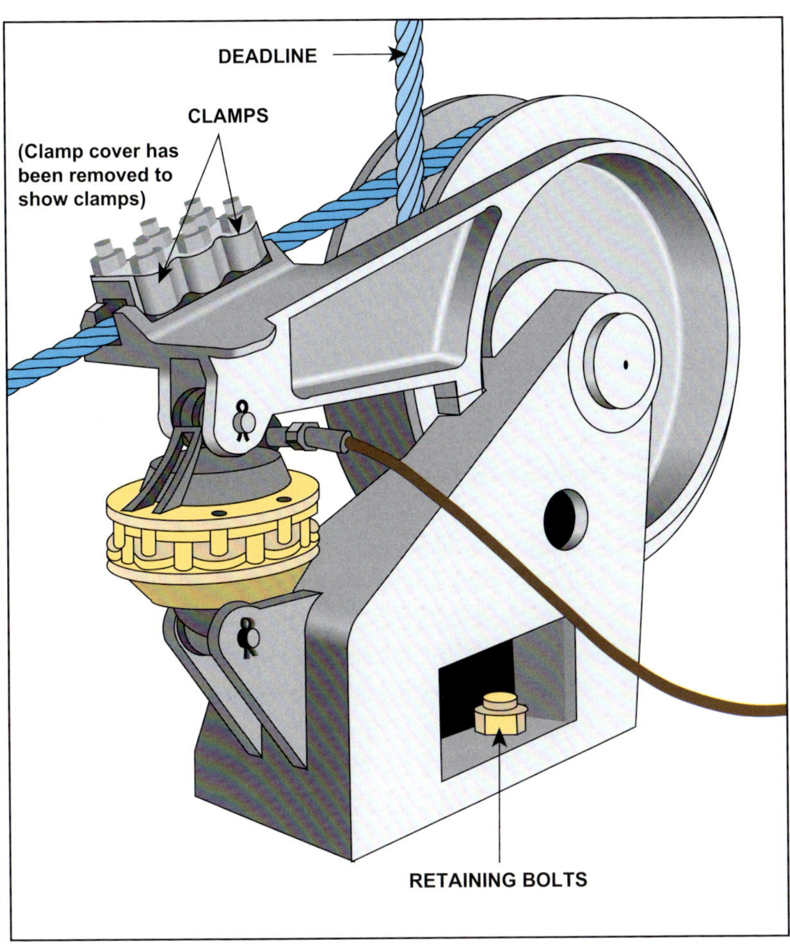

Figure 44. Deadline anchor

Inspect drilling line frequently for damage, replace it when it is worn, and handle it carefully to avoid damage.

Drilling line is critical to the drilling operation. It supports the entire weight of the drill string, traveling block, swivel, and hook. It should be inspected frequently for broken or damaged wires. Reserve line, if any, should be stored on a reel, and the line should be strung on suitable supports in order to keep it off the ground and to prevent kinks or other damage. Most contractors follow a slip-and-cut program based on ton-miles (megajoules) logged to assure even wear on the hoisting lines.

Drilling line is expensive and should be treated carefully, especially when slips and cutoffs are made. When the line is slipped on the anchor, use a wooden braking device, not a metal one, to protect the line. The traveling block should be lowered to the floor before unspooling line to be cut off. Cutting wireline must be done carefully to prevent unraveling (wickers). The line should be seized on both sides of the planned cut. Seizing is done with soft wire, not tape (fig. 45). A mechanical or hydraulic rope cutter is preferred over a hammer and chisel to make the cut.

On the drawworks end, the drilling line must be evenly spooled on the drum and secured to it by a clamp whose strength is equal to or greater than the breaking strength of the line.

Turnback rollers on both sides of the drum reduce scrubbing of the line where new layers are started. Hoisting line should not abrade against a mast, derrick member, or equipment. Line spoolers or stabilizers should be guarded, supported, and fixed securely in place. When spooling line onto the drum, care must be taken to avoid kinks and damage. When a worn line is replaced with another line, use a swivel connector (snake) to prevent transferring twists from one line to the other. Wirelines should not be struck or pried with any metal object. If a wrap must be moved over or crowded on the spool, use a wooden block between the line and the hammer; do not use a pry bar.

Sandlines, which are used for swabbing, tool runs, and surveys, should be spooled carefully to ensure level winding on the reel. The reel should have a proper line-spooling device. If a level winding device is not present, a manual guide with a wooden or rubber-covered roller may be used. Do not stand in front of the drum to guide the line onto the reel. Stand back from the rotary when a sandline or other cable is being run in the hole because slack loops may form and entangle anyone nearby.

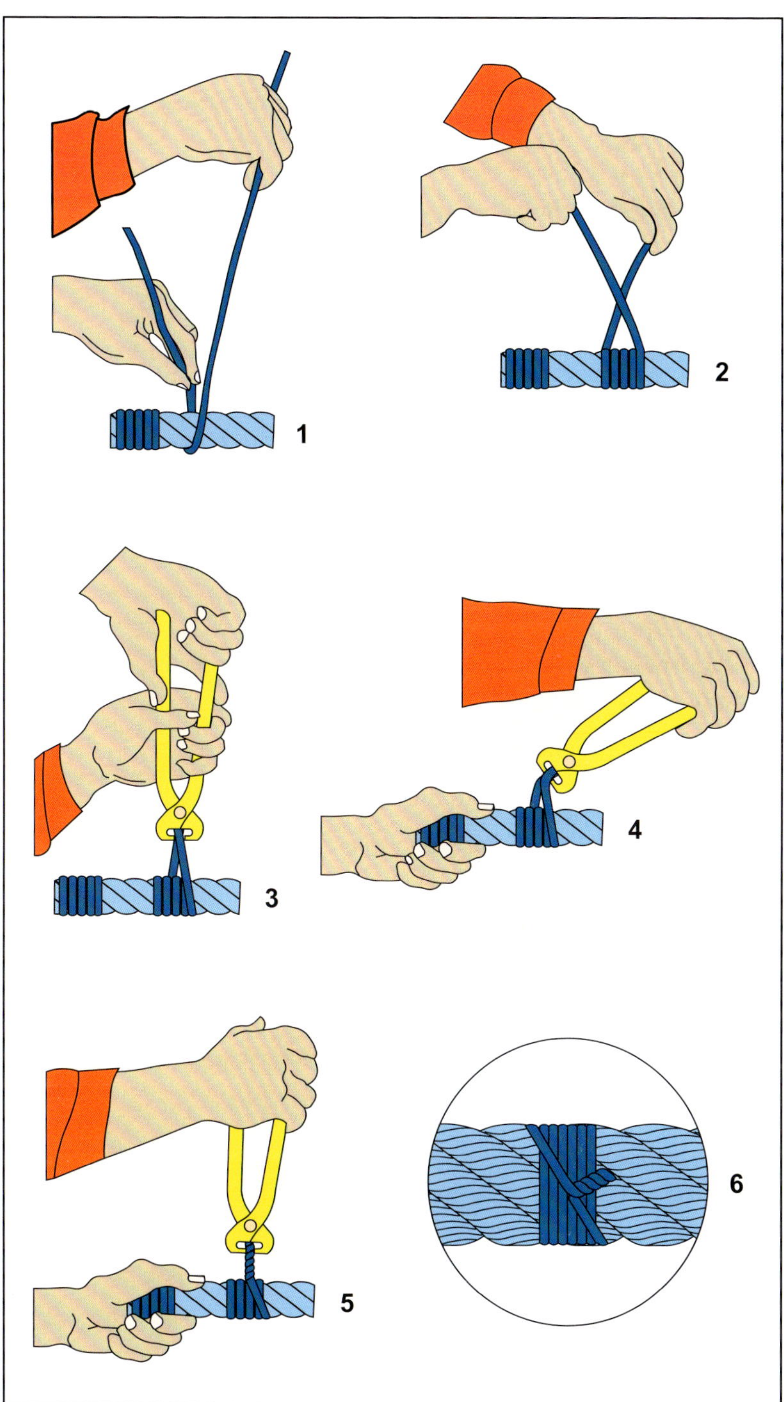

Figure 45.   Seizing of wireline rope

Slips

Slips perform a vital function. They support the full weight of the drill string during trips and connections. Worn slips could permit the pipe to drop, damaging the drill string and injuring crewmembers. They should be checked frequently for wear, cracks, or damage to the pins, dies, keys, handles, and body. Slips should not be modified. Grease the tapered sides of drill pipe slips to facilitate removal. Handles should work easily; lubricate as needed. Drill pipe slip handles should not project beyond the inner edge of the metal step around the top outer edge of the rotary table when the slips are set into position to hold the pipe. Choosing the right-size slip for the pipe to be handled is critical. Misfitting slips can gouge, scar, pinch, or otherwise damage the pipe. Pipe slips and collar slips are not interchangeable; use the proper type.

Air- or hydraulic-powered slips are safer than manual slips and less apt to damage the drill pipe (fig. 46). They eliminate the physical exertion of pulling and setting the slips when tripping the drill pipe. Be sure the power-input line is disconnected before working on them.

Floorhands working with manual slips should observe the following general procedures:

- Use the handles.

- At least two people should pull and set the slips (fig. 47).

- Use proper lifting techniques.

- Handles should be grasped with the palm up toward the pipe. If the back of the hand is up and toward the pipe, it may be more difficult to release the handle as the slip sets in place.

- Do not attempt to kick the slips into place.

- Keep hands and feet, as well as chains and ropes, away from slip handles when the rotary is in motion.

- Set two-piece slips at the same level to avoid damage to the pipe.

- Do not set the slips on the tool joint box. The results may be dropped pipe or damage to the slips or pipe.

- Do not let the slips ride the pipe while it is being pulled. This action will cause excessive die wear on the pipe, and the slips could flip out of the rotary, endangering anyone nearby.

- Do not set the slips while the pipe is still being lowered into the hole. If slips catch the moving pipe, the result may be damaged or dropped pipe.

Figure 46. Power rotary slips

Figure 47.   Crewmembers set slips around the drill pipe.

Pipe Tongs

Power tongs reduce strain that can lead to injury. However, only employees trained in their use should operate them.

Treat tongs with respect. Many rig injuries occur as a result of using tongs in the wrong body position or in other unsafe ways. Always use the tong handles to maneuver the tongs about.

Tongs should be inspected and lubricated prior to every trip. The dies should be sharp and properly pinned. If worn, they should be replaced as a set. Die drivers should be used to remove tong dies. Wear personal protective equipment when changing tong dies.

Tongs should be attached to counterbalance weights located in the mast or derrick. Weights may be suspended either above or below the rig floor but must be guarded to prevent worker contact with the weights or lines. Makeshift weights such as tools and bits should not be added to the counterbalance weights. Tongs must have safety lines attached to anchor (back up) posts which, in turn, are secured to the substructure, not to a mast or a derrick member or derrick leg. The safety line should be a wire line whose strength equals or exceeds the breaking strength of the tong's makeup or breakout line. Both ends of the safety lines should have three clamps properly and tightly attached. Line ends should be seized or welded to contain loose ends (wickers). Spliced eyes on the ends of safety lines are acceptable if properly done. The safety lines should allow full movement of the tongs but be short enough to prevent the tongs from rotating completely around the drill pipe.

Crewmembers must not stand between or within the arc of the two tongs while pipe is being made up or broken out. The driller must be sure the floorhands are clear of the tong handles before power is applied because a safety line may break or the automatic cathead may repeat its action, causing the tongs to rotate with the pipe. When excessive pull is needed to break a tight joint, all floor workers should move away from the rotary and out of the way of the tongs before heavy torque is applied.

Two tongs should be used during makeup and breakout (fig. 48). One set of tongs should never be used to break out a joint because the drill stem may rotate in the slips and damage the drill pipe. The rotary clutch should not be engaged if tongs are attached to the pipe or anything else in the rotary, with two exceptions:

- Tongs may be used to hold a joint or stand of pipe as it is spun out after the connection has been broken by the use of two tongs.

- The rotary may be turned slightly if the backup tongs are so tight after a joint was broken that they will not release.

Figure 48. Crewmembers use tongs to tighten one drill pipe joint to another.

Many rigs are equipped with air-powered spinning tongs. Power tongs should be rigged with safety lines similarly to regular tongs unless they are equipped with backup jaws. Whether spinning tongs are making up or spinning out, they should have safety lines attached to hold backup torque.

Air-powered spin-up tongs are safer to use than a spinning chain; however, only trained and designated personnel should operate power tongs. Disconnect the power input pressure line before doing any repairs or maintenance. Floor workers can run pipe by using spin-up tongs and air slips, and then use regular tongs to apply final torque to the connection.

Spinning Chains

Figure 49. Running pipe with a spinning chain

No one should operate a spinning chain without being thoroughly instructed in its use. Today, many contractors have eliminated spinning chains in favor of spinning wrenches or tongs. However, a spinning chain may still be used as a backup in case the spinning wrench fails. If the rig uses a spinning chain, the following procedures are presented as possible methods of use. Chains should be of a proper length, without a surplus amount, and properly maintained (fig. 49). The spinning chain should be connected to the cathead chain by a connector link of equal strength to the lighter chain. If a connector link is used, it should be located where it will not be in contact with the rotating pipe. The tail end of the chain should have an 8- to 12-inch (203- to 305-millimetre) length of manila or nylon rope. The rope provides a handhold on the end of the chain that reduces the possibility of hand or finger injuries.

The driller should be certain the spinning chain is positioned on the pipe and floor so that it cannot entangle the floorhand handling the chain. Both the driller and the floorhand handling the chain should be careful to assure that the chain is not on the threads of a tool joint because the pin threads and box shoulders can be damaged if the chain becomes lodged in the tool joint connection. The spinning chain operator should not slip the chain while the chain is under high pressure because this can scratch or scar the pipe above the tool joint.

Spinning chains should be safely stored when not in use. The chain should not be in contact with the rotary or wrapped around the pipe in the mousehole while the rotary is turning.

The driller should be protected from contact with the spinning chain by a roller-guard post. The post is located so that the chain moves freely but safely past the driller. Guidepost rollers should be mounted on lubricated bearings and move freely to reduce friction between post and chain. Spinning chain catheads are adjustable. They should be adjusted to provide proper torque on the pipe joint and for quick, free release when disengaged.

Elevators

Elevators should be maintained in good condition and inspected before each trip. Latches, latch springs, hinge pins, and elevator shoulders should be checked and lubricated as needed. Drill pipe elevator ear locks should be fitted with the proper-size steel bolts. Elevator links (bails) require little maintenance except occasional lubrication of the surfaces between the links and link-arms. Links should be checked for excessive wear by measuring and comparing the arm dimension to its dimension at the link eye.

The proper size and type of elevators must be used for the drill pipe, collars, casing, or tubing being handled. Bottleneck elevators bored for drill pipe tool joints should not be used for drill collars, and vice versa.

Elevators should not be used to transport personnel except in an emergency. Use only the elevator horns or handle links to control the elevator and avoid using the elevator links. Avoid grasping the elevators or bails near the link eyes (fig. 50). Heavier elevators can be handled more easily if a balancing strap or stabilizer is provided.

To avoid hand injuries, never place hands around the pipe until the elevator has been brought to a full stop and do not place hands on the pipe between the elevator and tool joint. Be certain the elevator door is securely latched before the pipe is picked up or swung toward the rotary. Elevators should be latched onto pipe lying in the V-door with the elevator door or opening placed on the upper side of the pipe and above the point where the pipe contacts the rig floor.

Use a pickup elevator with a cable sling to pick up casing. The sling should not be removed until the casing joint has been stabbed and rotated several turns. The casing elevator should be ready to be latched onto the casing as soon as the pickup elevator has been removed. Many operators prefer to use wire rope slings to pick up casing.

> Use only the horns or handle to control the elevator.

Figure 50.   Elevator link eyes

Cathead and Catline

Rig supervisors should instruct crewmembers in the proper use and maintenance of the cathead and catlines (fig. 51). No one is to use the cathead unless authorized by the driller or rig manager. The catline must be of hemp or manila material only. Do not use a frayed rope or a spliced rope if the splice will wrap onto the spool. A cathead spool should be smooth with no deep grooves. An antifouling device should be in use at all times. The cathead should be equipped with a catline divider to separate the first wrap of line from subsequent wraps. Maintain the manufacturer's tolerance between divider and cathead spool, but never exceed ⅜-inch (9 millimetres). Repair or replace damaged or grooved catheads and worn dividers. There should be nothing projecting from the moving parts that could catch on the operator's clothing. If the cathead shaft extends beyond the cathead spool, the shaft end and locking device should be covered with a smooth plate to prevent the rope from catching and winding around the shaft.

Figure 51. Cathead spool and catline

An automatic cathead (see fig. 51) is subject to grabbing or dragging and should be inspected regularly for proper adjustment. Wire ropes should be stronger than any extension piece fastened to them. Wire rope ends should be seized and clamped as recommended in Table 1 (page 71).

Important safety rules for cathead and catline operations include:

- Never use the cathead unless the driller or a qualified assistant is at the control console to shut off the power (fig. 52).

- Never add more lines than necessary on the cathead.

Do not use the cathead unless there is someone at the driller's controls who can turn off the power.

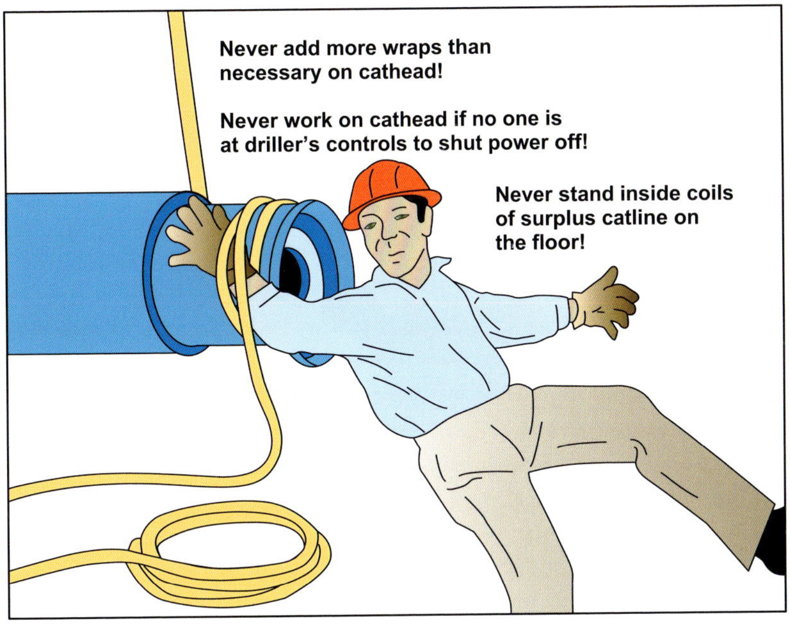

Figure 52. Possible consequences of not heeding safe practices in using friction cathead

63

- Keep legs and feet away from the surplus line coiled on the floor (fig. 53).

- Keep both hands on the line but never wrap the catline around your hand or arm.

- Do not leave an idle line on the cathead unattended after the load is removed.

- Never use a wireline on the cathead.

- The catline should never be used to lift personnel.

- The cathead operator must keep eyes on the moving load.

- No one should be beneath a suspended load.

Figure 53. Additional safety tips for using a friction cathead

The driller must watch all catline lifts. Do not block the driller's view of the load-lifting process. A designated signal operator should assist the cathead operator if more than one lifting operation is proceeding at one time and the driller's attention is being divided. A flagger or two-way communication device should be used when the cathead operator cannot see the object being raised or lowered.

The heavier the load is, the greater the number of turns are needed to support it; however, never use more turns than are necessary (see fig. 52). Extremely heavy loads subject the first turn of the rope to severe friction and heat that can weaken the fibers of the rope and cause a subsequent sudden failure. Always inspect the line for damage after an extra-heavy lift. Use the traveling block and hook equipped with a suitable sling to lift a load if more than 5 or 6 wraps would be needed on the cathead.

A cathead operator should use a swivel and safety hook attached to a lifting cap to pull pipe through the V-door. Another method of attaching the catline to a length of pipe is to use an endless sling on the pipe and then attach the catline hook to the sling.

The catline should not be used to lift equipment onto the derrick floor while the drill string is rotating because there is a danger of the line becoming entangled with the kelly.

Keep unused catline and rope neatly coiled in an out-of-the-way place to avoid a tripping hazard.

The driller must watch all lifts conducted with the catline.

Air Hoist

An air hoist demands the full attention the operator.

Air hoist pulling capacity depends on a number of factors other than simply the size of the unit. Air pressure, wireline size, and the number of turns on the drum affect the unit's pulling strength. Air hoists are rated from 1,000 pounds (445 decanewtons) up to 10,000 pounds (4,450 decanewtons) on the larger units. Wirelines range from ½-inch to ¼-inch (13- to 19-millimetres). The lines, hooks, and tail chains should be inspected frequently for damage and proper rigging. All hoists should be equipped with a drum guard and a line guide to protect the operator. At least six turns should always be left on the spool to prevent the line from pulling out of the drum anchor while under a heavy load.

Personnel should be instructed and trained in the use of an air hoist (fig. 54). Do not attempt to lift a load if there is a possibility the hoist line could become entangled in a rotating drill string. Heavy pulls should not be attempted when the drum is more than half full because the line can squeeze and compress on the drum. The crew-member operating the hoist should not be distracted by other work until the load has been landed and the unit secured. The brake must be set and the operator must not leave the controls while a load is stopped in suspension.

Figure 54. Air hoist and alert operator

Use a tag line to guide the load. No one should be under the load or attempt to guide it from below. A flagger should assist the operator when the operator cannot clearly see the load being handled.

To summarize—

Members of a drilling crew

- The crew on all rigs includes a rig manager, driller, derrickhand, motorhand, and two to three floorhands.

- The crew on offshore rigs also includes a crane operator, mechanic, electrician, and several roustabouts.

Rules for working safely in the mast

- Use a counterbalanced climbing device to get to the derrickhand's station.

- Safely transport tools and exercise caution when using them aloft.

- Tie back all stands of pipe in the fingerboard.

General rules for working safely with drilling equipment

- Install all equipment with proper supports and anchors. Securely fasten components with ties, latches, and bolts.

- Install appropriate shielding.

- Wear appropriate personal protective equipment.

- Check additional safety equipment for soundness.

- Inspect equipment frequently and make repairs or replacements as needed.

- Do not attempt maintenance or repairs until equipment is stopped.

- When shields are removed, clear personnel from areas with moving parts before restarting equipment.

- Select parts that meet or exceed strength requirements.

- Operate equipment at a safe speed.

- Only allow authorized personnel to operate equipment.

- Avoid distracting crew members who are operating equipment or moving loads.

- Position chains and lines that must be on the floor in a way so that they do not entrain personnel.

Rules for working safely with the hoisting and rotating systems

- Instruct all personnel in how to shut down the systems in an emergency.

- Only install and raise the crown block during daylight.

- Support the traveling block while reeving the line.

- Handle drilling line in a way to avoid damaging it and slip the line to prevent excessive wear.

- When the rotary table is in operation, keep the area around it clear of personnel and tools.

Rules for working safely with other pieces of drilling equipment

- Select slips that are properly sized for the pipe and follow procedures for using manual slips.

- Resist using pipe tongs incorrectly; use two of them to make up and break out pipe. Attach them with safety lines to anchor posts (rather than to members or legs of the derrick).

- Use the proper size and type of elevators for the tubulars. Only handle elevators with the horns or handle links.

- Assist the driller in watching all catline lifts.

- Do not leave the controls of an air hoist while the load is in suspension.

Rigging Practices

Rig personnel handle rigging every day, so it is important to recognize and correct unsafe rigging. All cables and rigging parts should be of good quality, size, and strength to handle the expected load (fig. 55). Check the manufacturer's recommendations if in doubt. Splicing, socketing, and seizing of wire rope should be done by a qualified person. Eye splices must have the proper size thimbles to protect the line from sharp bends and abrasion. The U-bolt attachments on wire rope clips must be fastened to the rope with the U-bolt side on the dead, or shortened, end of the rope. Align the clips with U-bolts all on one side of the rope (lower right, fig. 55). Check and tighten the clip nuts after initial use and frequently thereafter.

All cables and rigging parts should be of sufficient quality, size, and strength to handle the expected load.

To maintain the integrity of cable and rigging parts, handle them correctly and replace them when they are worn.

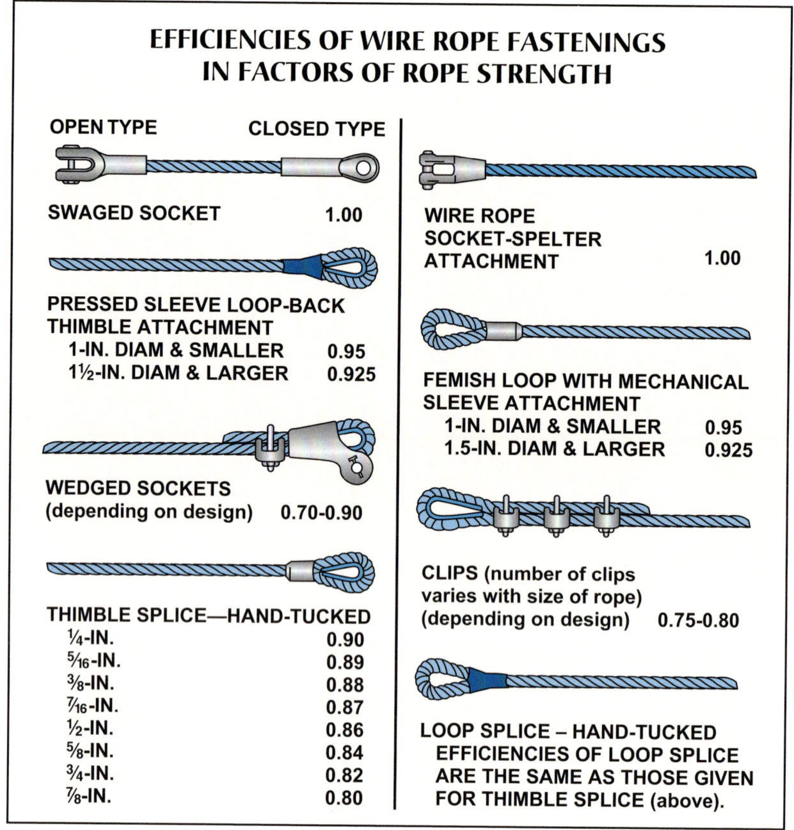

Figure 55. Wire rope fastenings

Table 1 shows the minimum number of clips, turn back length, and torque for the common wire rope sizes. The dead end on a wedge-socket fastening should be clamped as shown in figure 27.

Only certified lifting chains should be used for that purpose. Slings, fittings, and fastenings, should be inspected daily when in use. Replace any sling with defective or excessively worn parts. Place protective pads, wooden blocks, or spreaders between sharp surfaces and the sling cables to prevent sharp bends or kinks. Slings should never be choked in a spliced area.

Hook-on personnel should be instructed in accepted, safe rigging procedures.

Table 1
Minimum Number of Wire Rope Clips to be Used

Diameter of Rope (inches)	Number of Clips	Length of Rope Turned Back (inches)	Torque (foot-pounds)
⅛	2	3¼	4.5
3/16	2	3¾	7.5
¼	2	4¾	15
5/16	2	5¼	30
⅜	2	6½	45
7/16	2	7	65
½	3	11½	65
9/16	3	12	95
⅝	3	12	95
¾	4	18	130
⅞	4	19	225
1	5	26	225
1⅛	6	34	225
1¼	6	37	360
1⅜	7	44	360
1½	7	48	360
1⅝	7	51	430
1¾	7	53	590
2	8	71	750
2¼	8	73	750
2½	9	84	750
2¾	10	100	750
3	10	106	1,200

To summarize—

- Instruct personnel in safe rigging procedures.
- Follow manufacturers' recommendations for the proper quality, size, and strength of cables and rigging parts for the job.
- Only allow qualified personnel to splice, socket, and seize wire rope.
- Follow recommendations for fastening wire rope.
- Inspect rigging daily for defects or excessive wear.
- Where sling cables are in contact with sharp surfaces, use protective pads, wooden blocks, or spreaders.

Power Generation

▼
▼
▼

In this chapter:

- Safely starting rig engines
- Using lockout-tagout procedures during maintenance
- Alarms and automatic shutdown controls for emergencies
- Avoiding burns and exposure to exhaust
- Protecting the crew with shielding

Rig supervisors should instruct personnel in the proper operation and maintenance of the prime movers and auxiliary engines. Internal combustion engines of 30 horsepower or more should not be started by hand cranking. An external power source is usually available—one that provides compressed air or hydraulic pressure, an electrical motor, or a smaller gasoline engine. Do not start an idle engine with the multiple-engine compound by using the power from an engine that is already running. Engines or motors that start automatically should have a proper sign posted to warn of automatic operation, and no repairs to such machinery should be started unless a proper lockout-tagout procedure is followed. Positive lockout and tagout measures must be provided to ensure that an engine cannot be inadvertently started during repairs, inspection, or adjustment. Engines should be equipped with alarms and automatic shutdown controls that activate during an emergency or operational difficulties such as overheating, overspeeding, low oil pressure, or excessive vibration. Engine controls should be periodically checked and be capable of immediate shutdown of rig power in the event of personnel injury or equipment failure.

Engines

Engines should be equipped with alarms and automatic shutdown controls for emergencies.

Personnel must be protected from contact with moving or hot equipment. Make sure appropriate shielding and insulation are in place.

All exposed rotating parts such as radiator or cooling fans, belts, shafts, pulleys, couplings, or other moving parts must be provided with adequate shielding to prevent contact with personnel or with other equipment (fig. 56). A sign should be in place to warn of any damage to a shield or guard. Damaged guards should be repaired or replaced as soon as possible and prior to operation of the equipment.

Exhaust manifolds and piping should be constructed, installed, and maintained to prevent engine gases from leaking between the engine and the discharge line. Exhaust piping that could cause burns should be insulated or guarded. The discharge line should be directed away from the engine and work area. Spark-arresting devices should be installed in the exhaust system to prevent sparking and possible ignition of wellbore gases. The exhaust should be directed through a water spray during air or gas drilling. A similar arrangement should be in place during a drill stem test or other occasion when combustible gases are being vented.

Mufflers should be in place to reduce engine noise. Hearing protection should be worn around engines when noise levels might impair hearing.

Areas below engine mounts and skids should be kept clear of motor oil, filters, and debris.

Figure 56.   Guards in place on rig engine and generator

Compound

The power train for some rotary drilling rigs is a three-engine compound powering a drawworks and pumps (fig. 57). Multi-strand roller chain drives the drawworks. Either roller chains or V-belts may drive the pumps. All chains and belts in the engine compounds must be completely enclosed with a protective guard. Repairs or adjustments to the compound should be performed only after the clutches have been locked out and tagged out.

Nails or Allen wrenches should not be substituted for cotter pins on drawworks, rotary, or transmission drive chains.

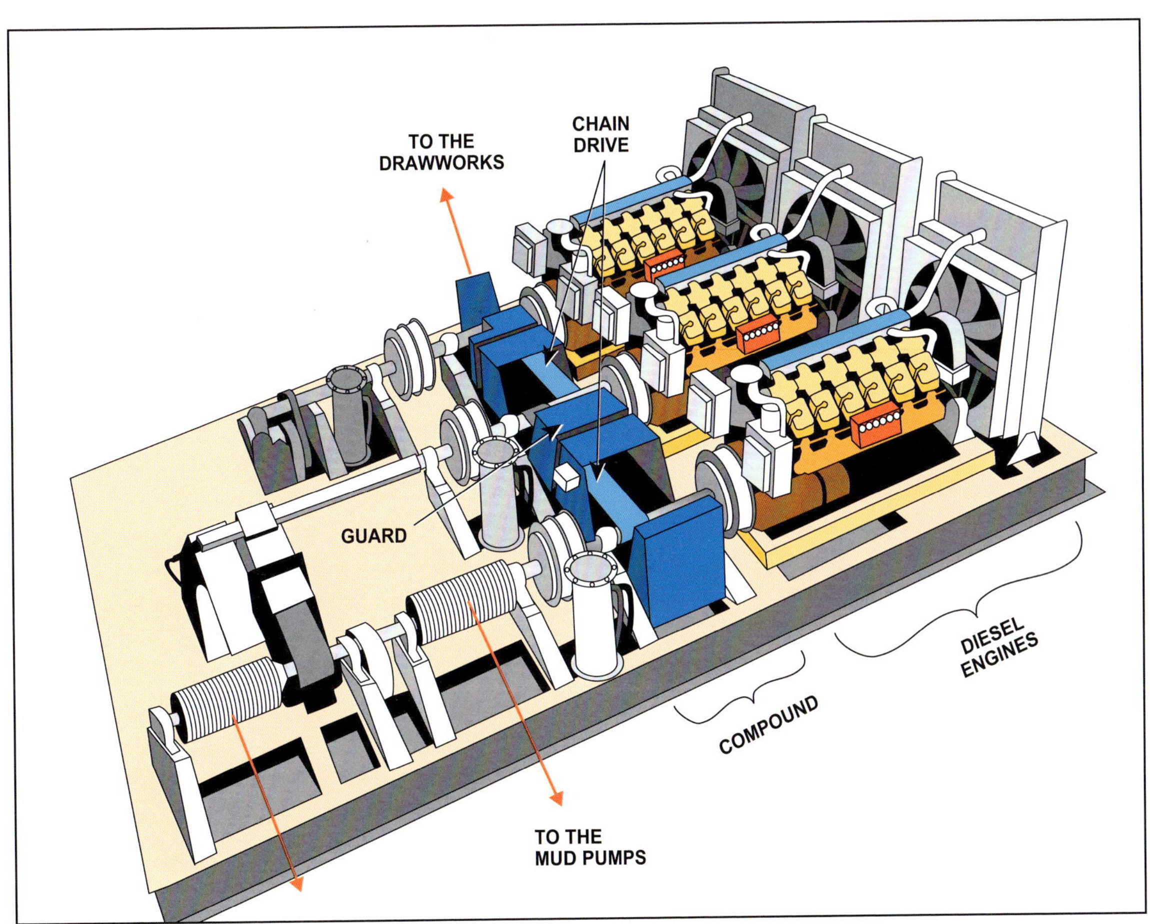

Figure 57. Three diesel engines and compound

To summarize—

- Instruct personnel in the operation and maintenance of the prime movers and the compound.

- Start engines with the proper power sources.

- Post warning signs and use a lockout-tagout procedure before inspecting, adjusting, or repairing components with moving parts.

- Check the integrity of engine controls, alarms, and automatic shutdown devices.

- Provide adequate shielding for moving parts and repair or replace damaged guards.

- Conduct exhaust away from enclosed spaces and insulate or guard hot exhaust pipes.

- Reduce the noise from engines with mufflers.

Mud Pump and Tank Safety

In this chapter:

- Installing pressure-relief valves on the mud tanks
- Safely performing maintenance on mud tanks
- Providing safe footing around mud pumps and tanks
- Safe procedures for mixing mud
- Personal protective equipment for mixing mud

Mud pumps provide a continuous supply of fluid under high pressure to the drill string (fig. 58). Caution should be used when working around high-pressure components. All pumps must be fitted with a pressure gauge. A plugged bit or inadvertently closed valve can create extreme pressures that will endanger all those nearby. The manufacturer provides pressure relief valves (pop-off valves) to protect the pump and discharge lines from failure. Some pumps are equipped with adjustable, automatic-reset relief valves. These types of valves allow continuous operation without resetting after pressure surges. Shear relief valves depend on shear pins sized to contain a desired pressure. Supervisors should assure that no oversize pins are used. Nothing should be done that would eliminate or restrict the operation of any safety device. Relief valves should be shielded to protect workers from flying parts in case the shear pin is broken by excess pressure.

A pressure relief valve of an approved type must be part of the discharge manifold and placed ahead of any valve on the line. Relief bypass outlets should be short, without bends, and directed away from personnel and equipment. Discharge lines should be securely anchored when run into the mud tanks. Bypass fluid should not be returned to the pump suction or wasted.

Mud Pumps

Personnel must not tamper with safety devices on mud pumps.

Keep the area around the mud pumps clean to avoid slip hazards.

Figure 58.   Single-acting triplex mud pumps

No maintenance is to be performed on the mud pumps without the driller's knowledge. The pump power source, whether air clutch, electrical, or V-belt, should be deactivated and controls locked out to prevent accidental operation of the pump when it is being repaired. The suction and discharge valves should be closed and all pressure released from the fluid end before attempting to remove the valves and plungers. Air lines to pump clutches should have a positive cutoff. Disconnect the air line if it has a petcock or ball valve because these types of valves can be accidentally bumped open.

The right tools are to be used for repairs. Hydraulic or mechanical liner, rod, or piston pullers should be available. Pump power should not be used to force liners or other components from the pump. Socket or box-end wrenches rather than pipe wrenches should be used. Sledge hammers should not be used on the pump or on wrenches unless the wrenches are designed for hammer use. Improper adjustments can cause damage and delays, so apply the correct torque to gland-packing nuts and other accessories. A pump should be primed before operation. A few strokes of the piston in a dry liner can ruin the piston.

Good housekeeping and drainage is needed around the pumps to eliminate slipping hazards created by oil or mud spills. Adequate lighting around the pumps is important.

Mud Tanks

Steel mud tanks (pits) must be provided with stairs, handrails, and walkways that permit safe access to the mud conditioning equipment—tanks, shale shaker, desander, and degasser (fig. 59). Handrails are needed on the inside of walkways, as well as outside, if the tanks are below the walkway level.

Figure 59. Mud tanks equipped with stairs, walkways, and guardrails

Tank Safety

The chemical mixing tank or barrel should have a paddle mixer and ample water supply for mixing caustic and other chemicals. Add caustic soda to water slowly to prevent splashing. The following safety items should be clean and readily available at the mixing hopper or other mud mixing location:

- Eye and face protection
- Air purifying respirator
- Rubber gloves and apron
- Eyewash station
- Warning signs

See the Chemical Hazards chapter (p. 95) for other caustic mixing precautions. Only designated employees should operate the mud guns.

Be sure mud gun unions and connections are tight. Arrange flow lines to eliminate tripping hazards. Mud tanks should be well lighted with approved explosion-proof fixtures. Place all electrical equipment where there will be no contact with fluids. The agitator's power source should be deactivated and locked out before anyone enters a mud pit.

Gas in the mud makes the mud lighter and well control more difficult. A degasser should be provided to remove the gas and vent it away from the shale shaker and pits (fig. 60).

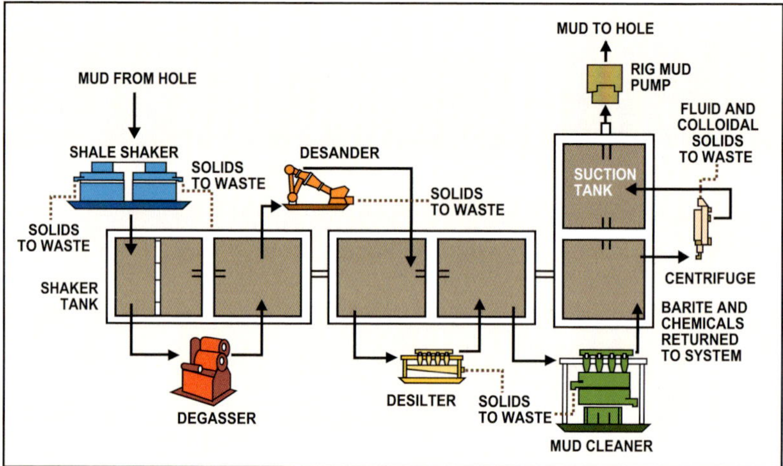

Figure 60. Equipment used to remove sand, silt, and gas from drilling mud

To summarize—

- Keep the area around the mud equipment well lit and free of slip and electrical hazards.
- Do not tamper with safety devices.
- Before performing maintenance, deactivate the power source and lock out the controls.
- Follow safety procedures for adding hazardous chemicals such as caustic to the mud.

Tubulars

The handling of tubular goods can be hazardous. "Never turn your back on moving pipe" is rule number one around the drilling rig. Moving pipe deserves your full attention.

Do not leave drill pipe in the mousehole during a trip. When a stand of drill pipe is hoisted off the floor, it should be held back to keep it from swinging. The derrickhand should help break the swing of the pipe as the stand is led across the floor. Pipe should be racked by pushing against the outer face to set it back; never put hands on the back side and keep feet away from the pipe as it is set down. Always keep hands on the outside of casing, drill joints, subs, or collars. Never place hands on top of any tubular held in the slips when another joint is being stabbed. The driller should always be aware of the stabber's hands when lowering the elevators to avoid catching them between the tubular and the elevators.

Rig Floor

Employees must always be aware of moving pipe around them.

Each drill collar sub should be tightly screwed into the collar so it does not back out when the collar is hoisted. Feet, hands, or knees should not be placed below the drill collar clamp while it is being tightened onto the drill collar. When drill collars are being pulled up in the mast or derrick, a drill collar clamp secured to a collar should not be hoisted into the mast or derrick above head height. Clamps can slip free, slide down the collar, and injure a floorhand if not removed. Drilling mud should be scraped off stabilizers and drill collars before they are lifted from the rotary and stacked. Stabilizers, wall scrapers, or pup joints should not be on the rotary table without being supported by a hoist line. Single-joint casing pickup elevators should be used and be equipped with a safety pin attached by a chain.

Do not place hands or feet under the core barrel opening when removing cores from the core barrel.

Do not place feet under the pipe when rabbiting the drill pipe. Face away from the pipe rack when the rabbit is dropped to avoid possible eye injury, even when safety glasses are in use.

Pipe Rack, Bins, and Catwalk

Figure 61. One type of pipe rack bin

To minimize hazards, keep pipe racks level and orderly. If the rack uses pins, make sure they are in place before pipe is laid on the rack (fig. 61). Each layer of pipe on the rack should be chocked with wooden blocks nailed to the stripping board. To move stubborn pipe use a pry bar, not your hands or feet. Do not walk on top of racked pipe, especially when it is wet or snow covered; place boards or plywood on the pipe to provide footing. Never stand between the pipe truck and the pipe rack, because someone on the other side of the truck may be unlatching a binder.

When rolling pipe on the rack, always roll the pipe from the ends. This places workers in a safe position in case the pipe gets out of control.

When moving pipe to the rig floor, there should be a flagger to watch and signal clearly to the hoist or catline operator. All employees should be clear of the V-door and catwalk prior to hoisting tubular goods (fig. 62). Make two wraps around the drill pipe, and fasten with the hook facing downward. Never attempt to lift more than three joints at one time up to the V-door. Keep the V-door safety chain in place when not moving pipe.

Take care when lifting pipe from pipe tubs to avoid hanging up on the edge of the tub. Tag lines provide better control when lifting pipe from a tub, rack, or truck.

Figure 62.   Casing being lifted from the catwalk up to the V-door

To summarize—

- Control the swinging of drill pipe as it is moved across the rig floor. Use tag lines to lift from a tub, rack, or truck.

- Clear crew from the V-door and catwalk.

- Use a flagger to signal the hoist operator.

- Do not move more than three stands of pipe at a time up to the V-door.

- Make up drill collar subs tightly to prevent them from backing out when the drill collars are hoisted.

- Keep hands and feet away from stands of pipe as they are set down, rolled, and made up.

- Keep pipe racks level and orderly.

- Place chocks between each layer of pipe on the rack.

- Place boards on top of racked pipe to provide safe footing.

- Roll pipes on the pipe rack from their ends.

- Do not stand between the pipe rack and the pipe truck.

Hazardous Energy

▼
▼
▼

All rig wiring should be insulated to prevent short circuits caused by weather, chemicals, and rough handling. Circuits must be built of standard, properly load-rated outdoor wiring and fixtures. Equipment repairs or cutting and splicing of electrical wiring should not be attempted by unauthorized personnel. Makeshift repairs are hazardous because an underrated element anywhere in the circuit can cause problems. If spliced, wiring should be equal to the original in strength and insulation. Wiring must be installed so that it is protected from abrasion, trampling, or burning by hot surfaces. Lead-in cables from the generators to the mast or derrick must be located and protected from damage when the rig is in operations or when heavy equipment is being moved. All guards on electrical equipment should be in place and in good repair. All motors, generators, equipment, lights, panels, and electric tools must have proper grounding. All auxiliary housing on location must be grounded.

Electrical Hazards

Lights are electrical equipment with glass that can shatter. Secure them carefully, use appropriate parts, and keep guards in place.

Mast or derrick wiring should not be strung on girts. All overhead lights in the mast or derrick should be equipped with safety fasteners and cables or straps to prevent them from falling if detached or broken. Lighting fixtures should be vapor- or explosion-proof with globes and gaskets in good condition. Lighting sockets should not have brass turnkeys or metallic pull chains.

Rig lighting should be equipped with reflectors or shields and directed toward the object being illuminated and away from the eyes of rig workers (fig. 63). The light at the derrickhand's station, for example, can be mounted below and behind the monkeyboard so that it allows both the driller and derrickhand to clearly see the block, hook, and elevators as they move past. No floodlight should be located in a way that blinds anyone on the derrick stairs, walkways, pumps, pits, catwalk, or pipe rack areas.

Hand lamps and flashlights used where gas may be present should be safety-approved models with explosion- and vapor-proof lenses. The lens should have a guard to prevent any breakage that would expose hot lighting elements to explosive gases. Explosive-proof lighting fixtures should be used to illuminate the wellbore if the substructure is enclosed.

Figure 63.   Rig lighting equipped with a reflector

All electrical receptacles should be the three-wire grounded type and be rated for the amperage and voltage of the circuit into which they are wired. Receptacles and switches should be vapor-proof, inspected regularly, and maintained in good condition. There should be sufficient receptacles placed around the rig to eliminate the need for numerous or long extension cords.

All portable electrical tools should have three-wire grounded cords and plugs or be double insulated. Electrical power tools should not have locking devices on their triggers, switches, or controls.

Electrical control panels should be properly constructed and grounded. The backs of the panels should be covered and securely fastened. Clean rubber or nonconductive mats should be placed in front of the panels to provide insulation for anyone standing on them. Wooden mats are less reliable because they allow moisture to collect between the mat and the floor. High voltage panels must be clearly marked "Danger: High Voltage."

Electric motors and generators must be grounded. Motors, generators, and other electrical equipment subject to arcing and located within 25 feet (8 metres) of the wellbore must be enclosed to prevent the entry of explosive gases.

Generators should be set and adjusted by an experienced electrician. They must be located away from all hazardous materials and where there is no possibility of becoming a source of fire. The alignment of generators is important for accessibility and maintenance. Each crew should have a member qualified to operate and repair the generators.

Transformer banks should be fenced and warning signs put in place. Metal fences must be properly grounded.

Employees operating or working around electrical equipment should observe the following rules:

- Do not take chances with electrical equipment or wiring. Be cautious. Nobody can tell a live circuit from a dead circuit by visual observation.

- If any equipment does not operate properly, notify your supervisor immediately. Do not experiment.

- Report broken wires, bulbs, bad connections, or other hazards to your supervisor and avoid contact with any exposed wiring (fig. 64).

- The breaker switch for any piece of electrical equipment should be disengaged and locked or tagged out before being repaired. (See the next chapter, which is on lockout-tagout procedures.)

Figure 64. Report broken or frayed wires.

Figure 65. Dry powder extinguisher

- Do not throw breaker switches "in" or "out" if a circuit is under load.

- Never open a blade switch while a motor is running because of the danger of arcing. A pushbutton or other approved stop switch is safer.

- Do not assume low-voltage electricity to be harmless. Low-voltage (110 volt) lighting circuits can be dangerous under certain conditions.

- Water and electricity do not mix! Electrical devices should not be operated where water hazards are present.

- Use water or other cleaning agents carefully around electrical equipment. Never wash down an electrical motor unless the power has been disconnected.

- Do not let water, oil, dirt, excessive dust, or trash accumulate around electrical equipment.

- Water, or any substance containing water, should not be used to extinguish an electrical fire. Use a dry powder (fig. 65) or CO_2 extinguisher labeled for Class-C fires.

- Know the location and classification of all fire extinguishers.

- Turn off electrical power tools before connecting or disconnecting the power supply.

- Do not exceed the manufacturer's recommendations when using power tools.

- Electrical power tools should not be used in a gaseous or explosive atmosphere unless specifically approved for such use.

- Never touch an electrical wire on or near the ground. A dangerous wire looks the same as a harmless one.

- Avoid stepping on or handling electrical equipment being repaired or modified.

- Turn off power before removing a fuse and use a fuse puller when replacing defective fuses.

- Install proper fuses. Do not use pennies or other pieces of metal behind fuse plugs. All fuse boxes should eventually be replaced with approved circuit breakers.

- Do not disturb ground wires (fig. 66); they are critical for protection from electrical shock and must be maintained in proper condition.

- Notify an electrician if an operation is to take place or a structure is to be built near an electrical line.

- Have someone observe and direct the operation when moving a high load near a power line. Take care not to contact high-voltage lines, guys, or light fixtures if working with a gin pole or forklift.

- Replace broken or burned out bulbs as soon as possible, but always turn off the circuit beforehand. Vapor-proof or explosion-proof units should be returned to the same condition when replaced or repaired.

- Repair junction boxes if they leak but do not drill drain holes in them.

- Use approved electrical service tools when making minor repairs.

- Extension cords should be maintained in good condition; discard any damaged ones. Disconnect cords before coiling for storage.

- Only authorized personnel should operate high-voltage equipment.

- Supplies of any kind stored around switchboards, fuse boxes, or electrical panels create fire hazards.

- Do not snub a winch line to a power pole. There is a danger of wires coming in contact when the strain is taken off the pole. The resulting short may part the wires causing the loose, hazardous wires to fall to the ground.

- Test all rubber goods before using them on electrical equipment.

- Be sure hands, clothes, and shoes are dry when handling energized electrical equipment.

- No jewelry, including rings, should be worn when working with electrical equipment.

Figure 66. Ground wires

Keep electrical equipment and the area around it dry and free of oil, dust, dirt, trash, and supplies.

An electrical flow through the human body to the ground is a possibility whenever an electrical current is present. When electricity flows through the body, the person may be badly burned or even killed. When someone undergoes electrical shock, immediately turn off the power causing the shock. If the power cannot be turned off, attempt to pull the victim away from the electrical source by using a dry stick, rope, or other nonconductive material. First aid for the victim, given by a trained rig medic or by another person following the procedures outlined in a first aid book, should begin immediately (see the First Aid chapter, page 145). The loss of a few minutes in starting artificial resuscitation may be the difference between life and death.

Lockout-Tagout

To control hazardous energy, it is recommended that drilling contractors develop a lockout-tagout program to be followed during the performance of repairs or maintenance on the rig (fig. 67). The purpose of the program is to prevent the unexpected start-up of equipment or release of stored energy that could cause injury. Just turning off the equipment is not sufficient; the energy source must be positively isolated to prevent accidental activation.

Figure 67. Supervisor inspecting lockout tag

Tagout is a method of marking equipment taken out of service. Under a tagout program, a tag would be placed on an air compressor that is out of service awaiting parts, for example.

Two or more tags are required to properly complete a tagout. Always tag:

- The device being repaired, cleaned, or inspected

- The main control, switch, or valve for the device

- The point where the energy source is isolated—the breaker or disconnect plug, for example

Lockout procedures involve an actual lock applied directly to the switch, circuit breaker, starter, valve, or other energy isolation mechanism after putting it in the off or safe position (fig. 68). A device is often placed over the energy isolation mechanism to hold it locked in the safe position. The driller, or other supervisor, controls access to the key(s).

All lockout-tagout devices must be durable and substantial. The person initiating or controlling the procedure should be identified on the tags or lockout device.

Examples of situations where lockout-tagout procedures are needed include, but are not limited to, repairs of electrical equipment, removal or replacement of guards, cleaning of jammed mechanisms, or any operation on equipment with moving parts that could entangle the worker's body.

Tagout procedures require for equipment that is out of service to be locked in the off or safe position and identified with tags.

Figure 68.   Label and lock on lockout device

To summarize—

- Select standard load-rated outdoor wiring and electrical fixtures.

- Only allow authorized personnel to repair electrical equipment, cut and splice wiring, or operate high-voltage equipment.

- Install guards on equipment and keep it in good repair.

- Ground all electrical equipment, control panels, and tools, as well as the auxiliary housing.

- Identify dangers such as high voltage with signs.

- Enclose all electrical equipment subject to arcing and located within 25 feet (8 metres) of the wellbore to prevent entry of gases.

- Enclose transformer banks in a fence and ground the fence.

- Do not allow water hazards to develop around electrical equipment.

- Know where extinguishers for electrical (Class C) fires are located.

- Develop a lockout-tagout program for repairs and maintenance.

- Disengage the breaker switch for all electrical equipment and lock/tag it out before repairing it.

- Before starting construction, identify electrical lines. Avoid touching them during construction or operations.

Confined Spaces Safety

In this chapter:

- Types of confined spaces
- Clearing a confined space with a hazardous atmosphere
- Preparing a confined space for the crew
- Personal protective equipment for workers
- Precautions for working in confined spaces

A confined space is considered to be any enclosure large enough for an employee to bodily enter and perform work but small enough to restrict the entry and exit of the employee (fig. 69). Confined spaces are not designed for continuous occupancy. Certain spaces are considered confined spaces if they have poor ventilation or a low oxygen level—the cellar and mud pits, for example. An authorized person such as the rig manager should test the space for oxygen deficiency, H_2S, or explosive gas before entry occurs.

Confined spaces are classified as:

- Nonpermit spaces
- Permit-required spaces

A nonpermit confined space does not contain atmospheric or other hazardous conditions with the potential to cause serious harm or death.

A permit-required space is one that:

- Contains or potentially contains a hazardous atmosphere
- Has the potential to engulf anyone who enters—with a cave-in, for example
- Has a floor or wall configuration that could trap or asphyxiate a person
- Contains any recognized, serious health hazard

Figure 69. Person being rescued from confined space

Authorized (permitted) personnel allowed to enter a confined space are specially trained and are equipped with SCBAs.

If a hazardous atmosphere is present, the confined space should be washed or steamed to clear the contaminants. Then it should be opened and ventilated for 24 hours, if possible, before personnel enter. Employers keep printed entry-permit forms on hand to control entry into permit space. The entry supervisor must prepare a confined-space permit form and post it at the confined-space entrance before work begins. The permit should specify:

- Atmosphere test levels
- Work to be done
- Special precautions
- Tools and safety equipment needed for the job
- Personnel involved in the operation, the supervisor, the persons doing the work, and the attendants

All employees involved in confined-space work as well as their supervisors are required to have special training. Workers entering the space must wear an approved self-contained breathing apparatus (SCBA) using Grade D air (fig. 70). Safety belts with life lines should be used by personnel entering a confined space. Use the buddy system: An attendant should remain outside holding the lifeline ready to assist and secure help in case of an emergency. No attendant or any rescuer should ever enter the hazardous space without SCBA.

Precautions should be taken with the air supply to see that blowers do not force exhaust fumes or other gases into the confined space. Never use oxygen for air supply because it creates a fire and explosion danger.

Power tools taken into confined spaces should be pneumatically driven, if possible. If electrical tools are used, cords and tools must be thoroughly inspected and grounded. Welding hoses must be inspected for leaks, and gas cylinders should be secured outside the confined space. When tanks are not in use, they should be cut off at the cylinder valve.

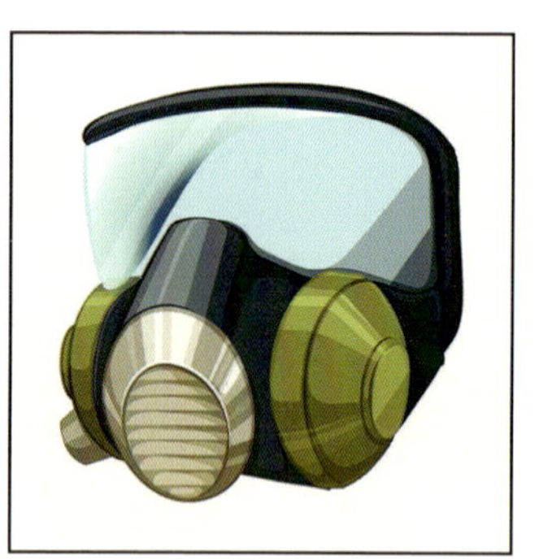

Figure 70.   SCBA

To summarize—

- Employers must control entry into confined spaces.
- To work in a confined space, workers must complete special training and wear the appropriate PPE.

Chemical Hazards

In this chapter:

- OSHA's regulations for chemical hazards in the workplace
- Communication to employees about chemical hazards
- Understanding warning labels and material safety data sheets
- Working safely with caustic solution

The regulations of OSHA's Hazard Communication Standard (HCS) require employers to inform workers of potential chemical hazards in the workplace. Chemicals pose physical or health hazards—or both. Physical hazards are dangers to the outside of the body. Health hazards cause damage inside the body, such as a stomach cramp or nausea. Health hazards may be immediate and short-lived or they may build up gradually over time with repeated exposure.

The employer should provide training by identifying and listing potentially hazardous materials. Warning labels and *safety data sheets* (*SDSes*) must be provided (fig. 71). You should be informed on detection methods, safe work procedures, and use of PPE.

Chemical Hazard Communication

Figure 71. Safety data sheets

Labels Labels warning of potential hazards should be in place and clearly legible (fig. 72). The label should give the name of the material, an appropriate warning of the hazard such as "corrosive" or "explosive," and information on the manufacturer or supplier.

- Never remove a label unless you replace it immediately with a new one.

- Do not assume a container is harmless just because it is not labeled.

- Never mix chemicals that are not labeled.

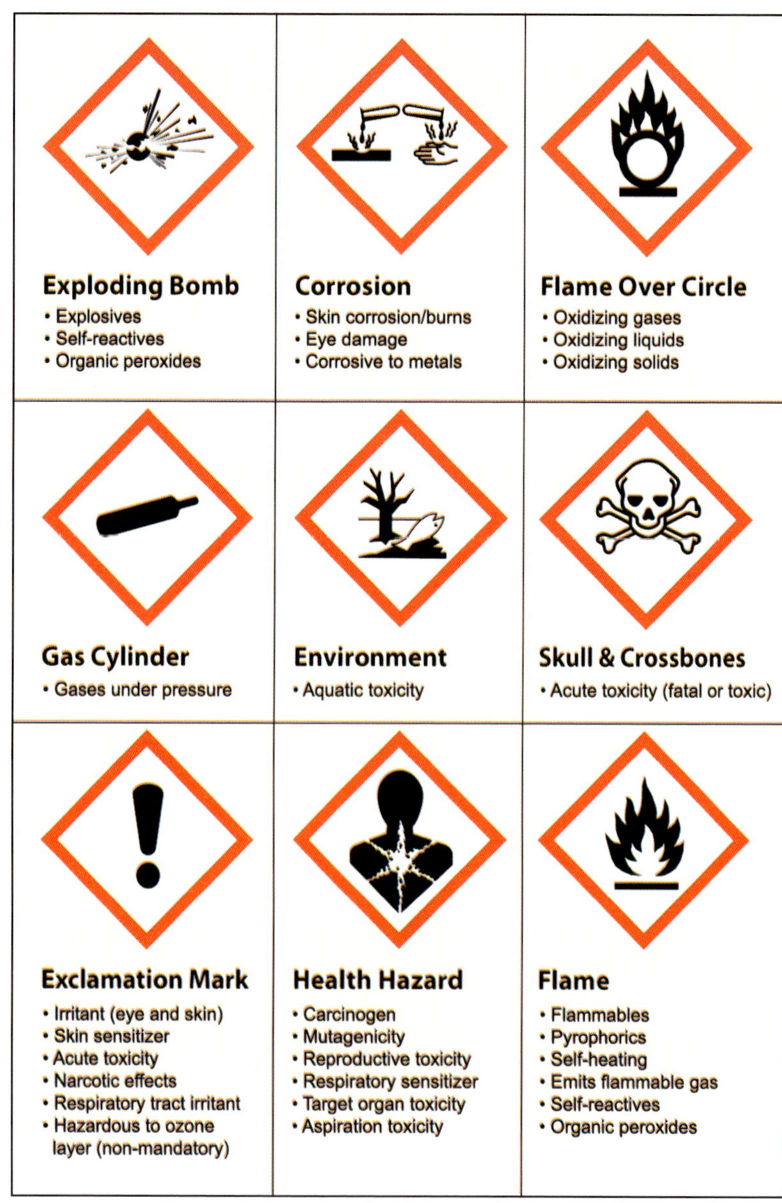

Figure 72. Pictograms used on warning labels

96

Safety Data Sheets (SDSes)

Beginning June 1, 2015, the Hazard Communication Standard will require employers to provide safety data sheets (SDSes) for potentially hazardous substances in the workplace (fig. 73). The new safety data sheets will replace the material safety data sheets (MSDSes) previously required. SDSes follow the recommendations of the Globally Harmonized System of Classification and Labelling of Chemicals (GHS) designed by the United Nations, so each SDS will resemble other ones around the world for the same product. In addition, the new sheets will provide more information than MSDSes did:

- Product information—name of the product and contact information for its manufacturer or distributor
- Hazard prevention—recommendations for handling, storage, exposure controls, and PPE
- Measures in case of exposure—first-aid and fire-fighting measures, measures for accidental release

Safety data sheets provide more information on chemical hazards than warning labels do.

OXI252
(disodiumflammy)
CAS #: 111-11-11xx

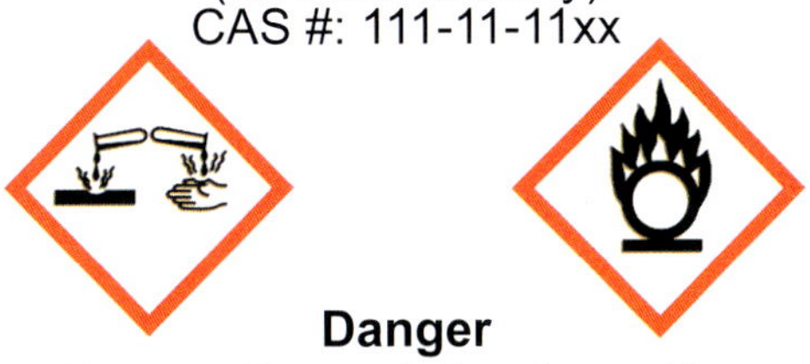

Danger
May cause fire or explosion; strong oxidizer
Causes severe skin burns and eye damage

Keep away from heat. Keep away from clothing and other combustible materials. Take any precaution to avoid mixing with combustibles. Wear protective neoprene gloves, safety goggles and face shield with chin guard. Wear fire/flame resistant clothing. Do not breathe dust or mists. Wash arms, hands and face thoroughly after handling. Store locked up. Dispose of contents and container in accordance with local, state and federal regulations.

First aid:
IF ON SKIN (or hair) or clothing: Rinse immediately contaminated clothing and skin with plenty of water before removing clothes. Wash contaminated clothing before reuse.
IF IN EYES: Rinse cautiously with water for several minutes. Remove contact lenses, if present and easy to do. Continue rinsing.
IF INHALED: Remove person to fresh air and keep comfortable for breathing.
IF SWALLOWED: Rinse mouth. Do NOT induce vomiting. Immediately call poison center.
Specific Treatment: Treat with doctor-prescribed burn cream.

Fire:
In case of fire: Use water spray.
In case of major fire and large quantities: Evacuate area. Fight fire remotely due to the risk of explosion.

Great Chemical Company, 55 Main Street, Anywhere, CT 064XX　　Telephone (888) 777-8888

Source: OSHA

Figure 73. Safety data sheet

Caustic Solution Safety

Caustic is a corrosive material that will cause a chemical burn when in contact with human flesh. (For more information on caustic, read its SDS). Certain precautions are therefore recommended:

- Alert other people before starting the mixing process. Warning signs should be in place.
- Have a quick-drench water hose and eyewash station (fig. 74) readily available.
- Put on safety goggles or a face shield, along with a chemical apron and gloves. Make sure your pants extend over your boots and are free of holes.
- When mixing, fill the barrel to within 6 inches (15 centimetres) of the top with cold water. Do not use warm or hot water; this speeds up the chemical reaction.
- Never dump the caustic into the barrel; slowly add it while stirring gently and avoid splashing.
- Do not mix caustic soda and paraformaldehyde together.
- Stand upwind with the wind at your back when adding caustic and make sure no one is working downwind from you.

Figure 74. Eye wash station

To summarize—

Chemical hazard warning labels

- Include the name of the material, a warning of the hazard, and information on the supplier
- Should not be removed

Safety data sheets (SDSes)

- Include the same information as a warning label
- Also include information on exposure limits and first aid, as well as directions for handling and storage

Rules for safely using caustic solution

- Warn others before mixing; post appropriate signs.
- Wear PPE, including gloves and goggles.
- Follow procedures for safe mixing.

Well Control

In this chapter:

- The danger posed by blowouts
- Main pieces of well-control equipment
- Safely installing and operating the blowout preventers
- Detecting a well kick and circulating it out

While drilling any hole for oil or gas, the crew can encounter abnormal pressures—ones that are either higher or lower than those expected. With abnormal pressures, comes the possibility a blowout, an uncontrolled flow of fluids into the atmosphere (fig. 75) or underground into lower-pressured zones. The danger to the crew, rig, and environment is obvious, as is the economic loss of a blowout.

A blowout preventer (BOP), choke manifold, and mud-gas separator provide a means to control pressures in the well.

Well-control equipment:
- Blowout preventer (BOP)
- Choke manifold
- Mud-gas separator

Figure 75. Rig blowout and fire caused by failure to control high formation pressure

Blowout Preventers

Blowout prevention equipment allows temporary, emergency control of formation pressure until heavier mud can be circulated into the hole to contain that pressure. Blowout prevention equipment is furnished by the contractor and varies according to the drilling hazards and pressures expected. All the blowout prevention components must be pressure rated to equal or exceed the pressures that might be encountered. The system must be kept in good working condition and an actuation test-performed on each round trip and prior to drilling out any string of casing (except conductor or drive pipe). Figure 76 shows a stack of blowout preventers designed for 10,000-psi (69-megapascals) operation. The BOP is fixed onto the wellhead flange attached to the surface casing; casing is designed, set, and cemented at a depth (usually below the freshwater level) to provide a solid anchor for the blowout preventers. During installation of the BOP, certain safety rules apply:

- Do not climb on the BOP or components until they are seated and bolted onto the wellhead flange or the flange of another BOP.
- Keep hands and feet from between the flanges while guiding the unit into place.
- Do not place hands or feet on top of the bolts while the unit is being lowered.

Figure 76.   Blowout preventer stack

- Use closed-socket hammer wrenches of the proper size to hammer up the BOP nuts and bolts. Select a sledgehammer of suitable size and weight for agility and accuracy. Use a rope to hold the wrench in place.

- Do not climb on the BOP while the stack is being pressure-tested.

Blowout prevention equipment may be actuated by either hydraulic or mechanical means. Small rigs may have hand-pump hydraulic controls, while larger rigs may employ power-driven pumps and accumulator vessels to provide fast closure on large preventers (see fig. 76). A master BOP control panel, usually located near the driller's console, should have each control clearly marked as to its exact function. Remote control systems should be installed so that failure of one set of controls does not affect the operation of the backup system. Designated employees should have an understanding of the BOP system and be able to operate it as needed. All new employees should be trained to operate the BOP equipment as soon as is practical. Practicing well-control procedures under simulated well kicks is an important aspect of crew training.

Choke manifold lines should be securely anchored to prevent movement resulting from pressure surges.

Accumulator units should be located at least 50 feet (15.2 metres) from the rig floor to avoid damage in the event of a blowout or fire. They should be housed and heated in freezing weather to assure free flow of hydraulic fluid. Antifreeze can be added to the hydraulic fluid if needed. Only qualified personnel should charge accumulators and high-pressure pulsation dampeners with nitrogen.

Time is critical when a well kick occurs. Most blowouts can be avoided if the crew is familiar with the early warning signs and knows the procedures to follow for controlling well pressure. Some of the preliminary events of a well kick are:

- Mud tank gain

- Increase in the flow of mud from the well

- Drilling break

- Decrease in circulating pressure

- Shows of oil, gas, or salt water in the mud

> Designated employees should be able to operate the BOP system, if needed.

Choke Manifolds

Several events may occur in sequence, such as a drilling break, flow of mud, influx of formation fluid in the tanks, and then a kick. Some indicators, such as salt water in the mud, may be obscure. Other events, such as pressure surges or swabbing, may be associated with pipe movement during a trip or connection and are routinely controlled by adjusting mud level, mud properties, or pipe movement.

Once a kick is detected, prompt action must be taken to keep the well under control. Four actions are necessary:

- Shut in the well by actuating the blowout preventors and chokes.

- Circulate out the formation fluids using the adjustable chokes in the choke manifold.

- Reduce the pump circulating pressure.

- Build mud weight.

The driller's drawworks controls should be tagged after a BOP ram is closed to alert the driller or the relief driller before pipe is moved.

Crewmembers should follow procedures for well control as directed by the operator's representative or other responsible individual. After shutting in, a safety meeting should be held to discuss the plan for controlling the well and where each crewmember is advised of the duties in the planned procedures. Control procedures may include some or all of the following:

- Target mud weight

- Choke size to set for the well

- Back pressure to hold on the choke

- Circulation rate to maintain while killing the well

Mud-Gas Separators

As noted previously, a choke manifold is used to control well pressure when circulating out after a well kick (fig. 77). A mud-gas separator is often used to handle gas kicks, as well (fig. 78). A mud-gas separator is a large vertical cylindrical vessel used to remove and vent large amounts of gas from the mud while the hole is closed in and the gas kick is being circulated out through the choke manifold.

Figure 77.   High-pressure choke manifold for well-pressure control

Figure 78.   Mud–gas separator

103

To summarize—

Well-control equipment

- Blowout preventer (BOP)
- Choke manifold
- Mud-gas separator

Rules for safely using a blowout preventer

- Test-perform the components on each round trip.
- Do not climb on the components.
- Watch hands and feet while the unit is being moved during installation.
- Install backup remote controls.
- Practice well-control procedures under simulated kicks.

Rules for safely using a choke manifold

- Install accumulator units at least 50 feet (15.2 metres) from the rig floor.
- Protect components from cold.

Symptoms of a kick

- Mud tank gain
- Drilling break
- Decrease in the circulating pressure
- Shows of oil, gas, or salt water in the mud

Procedure for circulating out a kick

- Shut in the well with the blowout preventers and chokes.
- Conduct a planning meeting.
- Circulate out the formation fluids with the choke.
- Circulate in heaver mud at lower pump pressure.

Well Servicing Safety

In this chapter:

- Planning for safe well servicing with third-party firms
- Preventing blowouts and fires during well servicing
- Handling casing and staying clear during cementing
- Preventing the accidental discharge of the perforating gun

Operations such as logging, drill stem testing, cementing, perforating, and fracture treating are done by third-party well service firms that are usually under contract to the operator. The service companies have their own safety rules that must be observed by rig employees, but it is the operator's responsibility to ensure that the service company's operations do not endanger the drilling operation or personnel. Service company personnel make up and handle their own equipment with assistance from the crew only as needed.

The service company supervisor, operator's representative, rig manager, and crew should conduct a safety planning session before any well service operation begins. Among points to consider are:

- Site hazards
- Hazards of the service operation
- Proper deployment of equipment and people that ensures safety and fire prevention
- Rig or equipment conditions that may affect the operation

Service company operations need a clear approach for placing equipment and space for vehicle parking. Other concerns include having a well-lighted and clean working area, obtaining pressure-testing records of wellhead equipment, operating blowout preventer controls, locating fire extinguishers and other emergency equipment, assigning personal protective equipment to the rig crew, posting signs, and identifying any unusual hazards present in the area.

Drill Stem Testing

Drill stem testing brings flammable liquids to the surface, so precautions are taken to assure the safety of the crew.

An open-hole drill stem test (DST) can provide critical reservoir information affecting the decision to complete or abandon a well. The test tool (fig. 79) isolates and opens a selected formation interval to the surface, thereby increasing the hazards to rig and crew. Most often, the tested interval is based on a drilling break and mud log or drilling shows. The following recommendations can help minimize the dangers involved in the test procedure, but specific conditions such as temperature and wind direction may dictate additional precautions. Most operators restrict drill stem tests to daylight hours because of the danger from flammable fluids coming to the surface. Some states have regulations prohibiting nighttime DSTs.

A meeting of all involved personnel should be held prior to running the test. It is important for everyone on site (operator, contractor, service company tester, rig workers, and others) to know how the test will be conducted and be made aware of the precautionary measures to be taken. The meeting should clarify who is in charge during the test (usually the tester or operator representative), individual duties, safety matters, lighting and smoking restrictions, and timing. Other matters discussed may include personal protective equipment, warning signs, and the confidentiality of test results. Breathing equipment may be needed and, if so, crewmembers should be briefed on its proper use.

Certain preliminary actions are usually taken prior to running a DST. The mud properties may need to be altered to better handle anticipated pressures and hole conditions. The mud may be circulated for a few hours to condition the hole and reduce the possibility of a stuck tool.

Fill-up lines must be installed on the drilling nipple above the blowout preventer to keep the casing filled with mud and must be used for that purpose only. Kill lines need to be installed and used separately from fill-up lines. A test line should be laid to the reserve pit and anchored securely. If the test recovery fluid is to be flared, two pilot flares should be in place to assure ignition is achieved under either high- or low-discharge conditions.

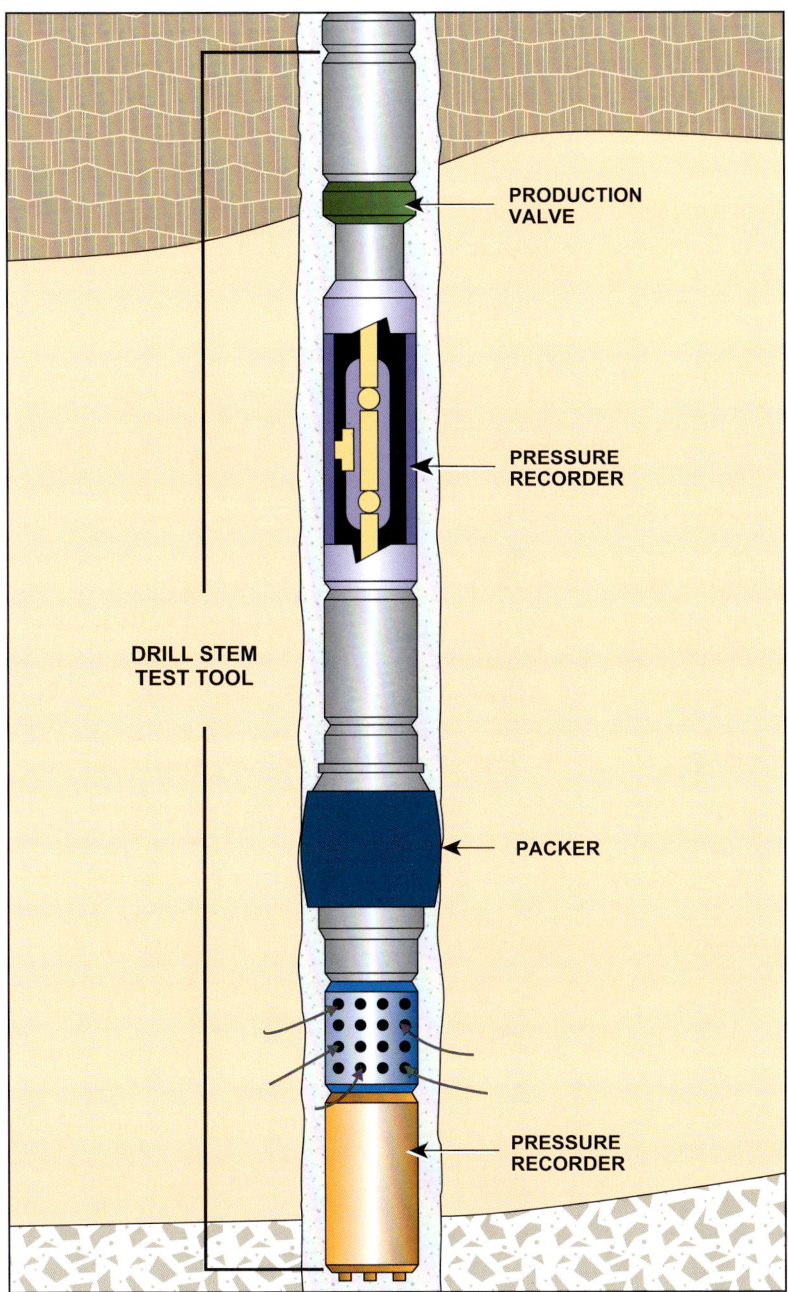

Figure 79.   Drill stem test tool

Blowout preventers should be checked for proper operation and pressure-tested prior to running the DST tool in or out of the hole.

The cellar should be drained, and blower fans should be in place to clear the air beneath the rig floor if there is any chance of hydrogen sulfide gas being recovered.

All potential ignition sources such as heating stoves, open fires, and rig lights should be extinguished before the test tool is opened. No engine should be running within 100 feet (30 metres) of the well unless equipped with a heat and spark-arresting device for the exhaust system.

Fluid volume in the casing should be monitored while the test tool is going in or coming out of the hole to assure that the well remains in hydrostatic balance. The mud level in the annulus should be checked regularly throughout the test to (1) ensure that the packers are holding, and (2) check for gas escaping from the test string, up-hole zones, or around the packers into the annulus.

The risk of fire or blowout is greatest when tripping the test string out of the hole; extra caution by all personnel is necessary during that time.

Open-Hole Wireline Services

Upon reaching total depth, the borehole must be evaluated to determine if formations capable of producing oil or gas have been penetrated. That's the whole point of drilling the hole. Electronic or radioactive instruments attached to a conductive wireline are lowered into the hole and information recorded as the tool is raised to the surface. The log data includes such critical information as rock type and depths, porosity, and fluid content. Electric logs are often supplemented by other wireline services such as sidewall cores and formation tests.

Wireline operations create potential well-control problems because the pipe is out of the hole and extraneous equipment is in the hole. Under these circumstances, mud level and condition should be closely monitored.

Wireline service company personnel handle their equipment on location. The rig crew assists as needed but generally stays out of the way. Wireline tools are made up on the catwalk, so that area should be clean and clear of obstacles. A rig helper may operate the air hoist when logging tools need lifting. A clear view and good coordination of signals are critical to safe operations.

Radioactive sources are used in logging tools. The service company is responsible for handling and maintaining such sources in safe condition. Warning labels or signs must be clearly posted.

Running Casing

After all information from the borehole has been analyzed, a decision must be made. Will the well produce commercial quantities of oil and gas? If the answer is positive, the operator orders casing, and well completion work starts.

The drilling rig is usually used to run production casing. A casing crew is contracted to supplement the rig crew and to supply certain specialized equipment. Teamwork between the rig crew and casing crew is vital because of the added people and high level of activity.

Rig components should be checked prior to running casing. A visual inspection of the hoisting equipment is important because of the heavy loads involved. The derrickhand should check the condition of the mud pump. The water pump should be fueled and working properly, and the water supply must be sufficient for running and cementing the casing. BOP rams need to be changed to fit the casing being run. If there is the possibility of hydrogen sulfide being present, blower fans should be installed to ventilate the cellar and area below the derrick floor. The drill pipe should be hoisted out of the hole before the casing equipment is rigged up.

The casing crew contractor supplies certain specialized equipment that includes heavy casing elevators, slips, and elevator links. A powered stabbing board may be furnished; if not, safely secured boards or a stand should be erected to allow the derrickhand to stab the shorter casing joints without overstretching or becoming off balance. Power casing tongs are safer, faster, and less fatiguing than a cathead-powered system. They also ensure that the proper torque is uniformly applied to each joint.

Casing should be handled carefully and safely when being moved from the rack to the derrick floor, picked up, stabbed, and made up (see the chapter on tubulars, page 81). A quick-release protector should be provided to prevent damage to the male threads.

Casing must be kept filled as it is run to prevent downhole collapse. An automatic fill-up shoe permits fluid to enter the casing as it is lowered.

Rig crewmembers should stay clear of the derrick floor when cementers are attaching the cementing head to the casing string, connecting hoses, or circulating cement. Cement service workers working from a stabbing board should wear safety belts and lifelines similar to those used by rig crewmembers.

When the casing has been landed, the rig flooring is removed so that the crown block can be used to lift and move the BOP stack. The blowout preventer stack should be unbolted and lifted over the casing using a proper-size bridle and spreader bar and then set on blocks. This frees the space around the casing so that the casing hanger and flange can be installed, the casing cut off, and the hanger seal-tested. Rig workers should stay clear while the casing hanger is pressure-tested by a qualified operator using proper equipment.

During well servicing, exercise caution while tripping out the drill string and monitor mud levels closely.

Perforating

Rig workers may assist the service crew in rigging up and rigging down, but not in actually perforating the casing.

After the production casing has been set and cemented, the casing and cement must be perforated to allow formation fluid to move into the well. The principal hazard in perforating operations is the possibility of accidental discharge of the perforating gun while on the surface. Rig workers may assist the service crew to rig up or rig down, but not in the actual perforating activities. Both rig and service personal should verify that all safety switches are in working order.

The preliminary safety meeting should emphasize these rules:

- All unnecessary personnel should keep away from the perforating operation.

- No smoking or open flames are allowed.

- All electrical equipment must be properly grounded.

- No repairs are allowed on the electrical system, including the power plant or generators, during perforating operations without the consent of the service company supervisor.

- All radios, transmitters, and radar within 500 feet (152.4 metres) of the rig must be turned off.

- Operations must be suspended in the event of an electrical or dust storm.

- All explosives will be cleaned up by service personnel.

To summarize—

General rules for safe well servicing operations

- Check equipment before beginning.

- Do not allow fire or electricity near the wellhead.

- Carefully monitor fluid levels during operations.

- Handle components carefully.

- Stay clear when the crew from the well servicing company is operating.

Field Welding and Cutting Safety

In this chapter:

- Using a qualified welder for cutting and welding
- Special precautions for drilling near the hole
- General precautions for welding and cutting
- Personal protective equipment for the crew to wear

Only a qualified, experienced welder should perform welding and cutting operations. A contract welder is usually employed for this work and assisted by the rig crew. Permission and a hot work permit must be obtained before welding operations start anywhere within 150 feet (46 metres) of the borehole. Field welding is not permitted on tongs, elevators, blowout preventers, or other heat-treated equipment.

Rig personnel who assist in welding operations should observe the following general precautions:

- Unless approved by the senior site supervisor, welding and cutting should not take place during high danger periods, such as when there are well-control problems or testing is taking place.

- The work area must be clean and clear of litter and combustibles.

- Rig workers should wear PPE, including goggles or a helmet (fig. 80), welders' gloves, and fall protection equipment.

- The work area should be well ventilated, and respirators should be made available if necessary.

During welding operations, personnel should have PPE and the proper permits, take the precautions recommended, and be ready to extinguish fires.

Figure 80.   Welders should wear proper protective gear.

- The welder should attach the ground lead directly to the metal being welded.

- Those working in a confined space should have a proper permit and follow special precautions. (See the Confined Spaces Safety chapter, page 93.)

- Precautions should be taken to protect those below when working overhead.

- Welding or cutting should not be performed on the pipe or casing on the racks.

- One or more persons should be assigned to fire-watch duty. A fire watcher should be familiar with the fire extinguishing equipment and have it readily at hand.

To summarize—

- Only allow a qualified welder to cut and weld.
- Before work begins, obtain proper permits and permission, if necessary.
- Do not weld or cut during dangerous operations such as testing.
- Wear appropriate personal protective equipment and keep respirators available.
- Keep the work area well ventilated.
- Take precautions to protect other workers in the area.
- Be ready to extinguish fires.

Fire Detection and Suppression

In this chapter:

- Training in fire prevention and suppression
- How fires burn and are extinguished
- Operation and care of fire suppression equipment
- How fire detection equipment works
- Personal fire safety equipment on a rig

lammable materials are all over a drilling site—oil and grease, natural gas, solvents, rubber hoses, cloth, and paper. Ignition sources are common, as well—lit cigarettes, welding torches, and sparks from motors, for example. So fire prevention, detection, and suppression are crucial to the safe operation of a drilling rig.

Everyone on a drilling rig should have training in fire prevention and take every precaution to prevent fires: Where you see a no smoking sign, for instance, don't smoke. Anyone servicing or operating equipment that involves sparks or flames must know when and how to work safely.

All persons on a drilling rig should know what to do if they see a fire and know exactly what to do and where to go when a fire alarm sounds. Everyone should know where the rig's fire extinguishers are and how to operate them. Especially offshore, all crewmembers depend on each other for safety in the event of a fire.

Everyone on a rig should be trained in fire prevention and know what to do when there is a fire.

Fire Prevention

Certain precautionary fire-prevention measures should be instilled in the crew:

- Smoking and nonsmoking areas should be clearly posted and strictly enforced.

- Rig heaters must be approved before use. In certain areas, all open lights or fires may be prohibited; those areas should be clearly posted.

- Open fires and any equipment operated with a flame must be located at least 100 feet (31 metres) upwind from the derrick floor. A fireguard should be on duty to keep burning trash from blowing around. Stake down flare lines and keep the burn pit a safe distance from the rig.

- Investigate all unusual odors, smoke, or gases. Determine the source and promptly report the problem. If the presence of flammable or toxic vapors is suspected, a qualified person should test the air with appropriate vapor testers. All ignition sources must be removed if the flammability limit has been exceeded.

- There should be an approved hot work permit filed before any welding and cutting operation begins. Such operations must cease in the event of a potentially hazardous situation, like a well kick or DST. A fire watch with an extinguisher should be set up when welding or cutting takes place.

- Matches and all other smoking material should be left in places designated safe for smoking. This restriction is especially important around a producing gas well.

- Oil, diesel fuel, and other flammables should not be allowed to collect around or under engines, cellars, or similar places. Keep all rig housing clean and orderly. Do not allow oily rags, clothes, or trash to accumulate and create the possibility of spontaneous combustion. Use designated containers.

- Spilling during refueling or fuel transfer should be avoided. Static electricity can ignite a fire; all fuel containers and equipment should be grounded or bonded.

- Flammable and combustible liquids should be stored in a proper container in designated, posted areas. Fuel and oil storage tanks should not be located uphill from the location or anywhere they might add to a fire at the rig. Label all tanks, barrels, or other containers for content: Knowing the content of vessels is helpful to anyone fighting a fire.

- Explosion-proof covers should be kept in place except during repairs.

- Electrical circuits must not be overloaded. Turn the current off before any repairs are attempted.

- Gasoline should not be used as a solvent or cleaning agent; use approved cleaning agents to clean equipment, parts, or tools. Do not use butane or propane for anything except as fuel.

- All liquefied petroleum lines, fittings, and valves should be leak proof. Test for leaks with soap and water, never with an open flame. Piping should be protected from damage and freezing by burial or other means.

To prevent fires, avoid exposing combustibles to flames, check lines for leaks, and investigate unusual odors.

Life Cycle of a Fire

Fire is a form of *oxidation*, a chemical process in which some substance combines with oxygen. The rusting of iron and the rotting of wood are forms of slow oxidation. Fire, or combustion, is rapid oxidation; that is, the burning substance combines with oxygen very quickly. Combustion gives off energy in the form of heat and light.

The Start of a Fire

All matter exists in one of three states—solid, liquid, or gas (vapor). The atoms or molecules of a solid are packed closely together, and those of a liquid are packed loosely. The molecules of a vapor are not packed at all; they move about freely. For a substance to burn, its molecules must be pretty well surrounded by oxygen molecules. The molecules of solids and liquids are too tightly packed to be surrounded. Therefore, only vapors can burn.

When a solid or liquid is heated, its molecules start to move rapidly. With enough heat, some molecules break away from the surface and form a vapor just above the surface. This vapor can now mix with oxygen in the air. If the temperature is hot enough to raise the vapor to its *ignition temperature*—the temperature at which a particular vapor burns—and if enough oxygen is present, the vapor burns.

Burning

Burning is the rapid oxidation of millions of vapor molecules. The molecules oxidize by breaking apart into individual atoms and recombining with oxygen into new molecules (fig. 81). This chemical process gives off energy in the form of heat and light.

The heat from a fire is *radiant heat*. Radiant heat is the same sort of energy that the sun gives off and that we feel as heat. It radiates, or

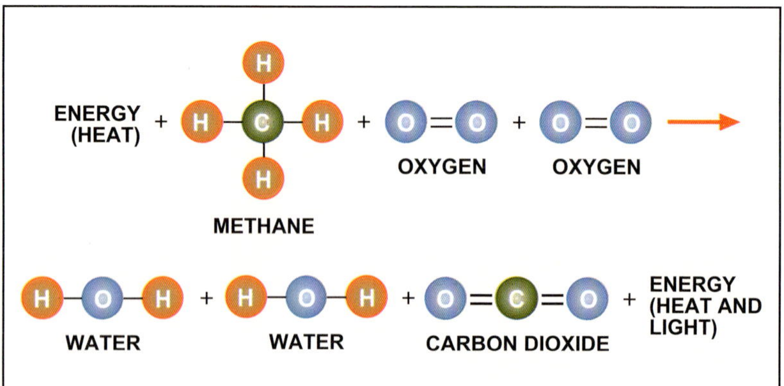

Figure 81. Molecules of a hydrocarbon, such as methane, combine with oxygen when heated. The reaction produces energy we can see and feel in the form of light and heat.

travels, in all directions, so part of it moves back to the burning solid or liquid (the fuel). The portion of the heat that moves back into the burning fuel is called radiation feedback (fig. 82). Radiation feedback causes the fuel to release more vapor and it raises the temperature of the vapor to its ignition temperature. At the same time, air, and thus oxygen, is drawn into the area where the flames and vapor meet. The result is that the newly formed vapor begins to burn, and the flames increase.

Figure 82. Radiation feedback is heat that travels back to the fuel from the flames.

Radiation feedback begins a chain reaction—the burning vapor produces heat, which releases and ignites more vapor. The additional vapor burns, producing more heat and releasing and igniting still more vapor, and so on (fig. 83). As long as fuel is available, the fire continues to grow and produce more flames.

Growing and Fading

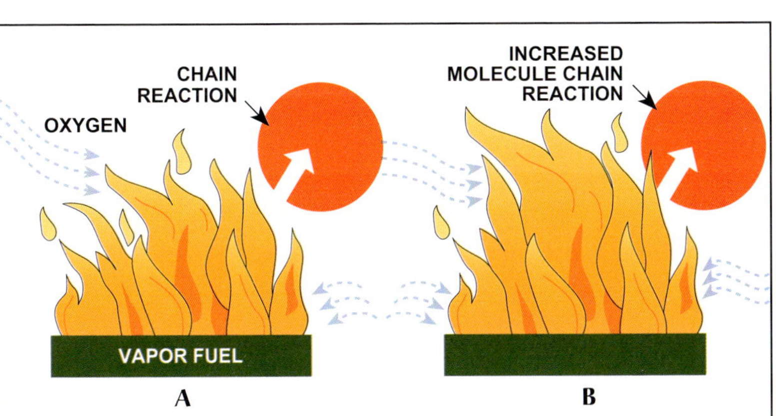

Figure 83. The chain reaction of burning. a) Vapor from heated fuel rises, mixes with air, and burns, producing enough heat to release more vapor and to draw in air to burn that vapor. b) As more vapor burns, the flames grow, producing more heat, releasing more vapor, and drawing in more air.

After a time, the amount of vapor released from the fuel reaches a maximum rate and begins to level off. The fire burns steadily and continues to burn until most of the fuel is gone. At this point, less vapor is available to oxidize, so less heat is produced. The chain reaction starts to break down. Still less vapor is released and less heat produced; less heat and flame, and the fire begins to die out. A solid fuel may leave an ash residue and continue to smolder for some time. A liquid fuel usually burns up completely.

Burning Gases

Gases burn more intensely than solids or liquids because they are already in the vapor state. All the energy of radiation feedback goes into raising the temperature of the vapor and igniting it. Gases burn without smoldering or leaving any residue. The size and intensity of a gas fire depends on the amount of fuel available—for example, a large tank of gas or a small canister.

Fire Triangle

Combustion requires three things: fuel, oxygen, and heat. The fire triangle (fig. 84) illustrates these requirements. It also illustrates two important rules of preventing and extinguishing fires:

- If any side of the fire triangle is missing, a fire cannot start.
- Remove any side of the fire triangle from an ongoing fire, and the fire will go out.

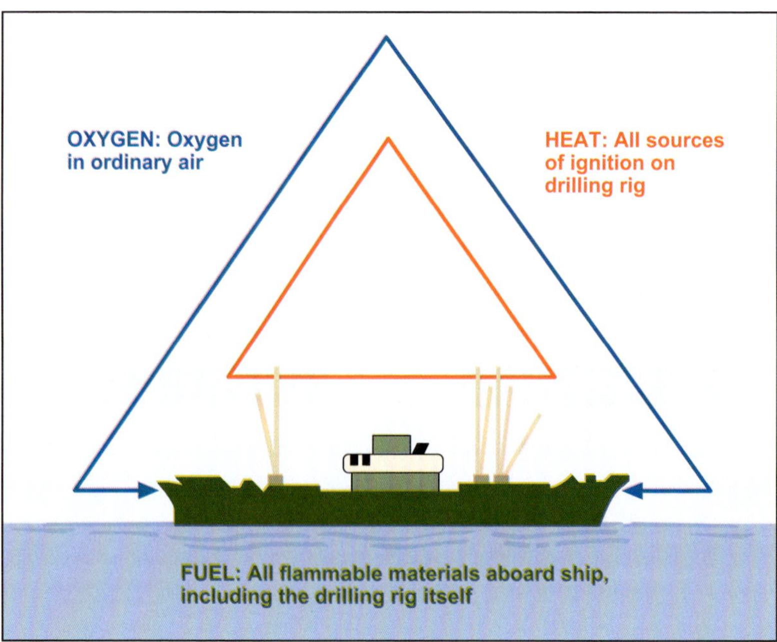

Figure 84. The fire triangle. Fuel, oxygen, and heat are necessary for combustion.

120

The purpose of fire suppression equipment is to remove the heat, the fuel, or the oxygen from a fire, or to interrupt the chain reaction of burning.

Extinguishing a Fire

The most common way to extinguish a fire is to remove the heat with water. Water absorbs heat very well, both from the fuel and from the radiation feedback. Drilling rigs on land and offshore have hoses for directing water to a fire.

Removing the Heat

One way to remove the fuel is to physically drag it away. With a solid fuel, this is usually impractical, although removing nearby fuel that is not yet burning is a good way to prevent the fire from spreading.

When the fuel is a liquid or gas coming from a line, shutting off the proper valve will remove the fuel source. Some equipment has safety valves that can be shut off by the pressure from a stream of water.

Removing the Fuel

Removing or reducing the oxygen puts out a fire by smothering it. One way to smother a fire is to flood the area of the fire with an inert gas such as carbon dioxide that displaces the oxygen. Inert gases do not react with either the fuel or oxygen. This method is difficult or impossible to use in an open area. Carbon dioxide (CO_2 for short) would quickly blow away from an open deck, for example. It is useful for fires in contained areas, such as a sealed compartment or room.

Another method of removing oxygen is to smother the fire with foam. Foam floats on top of a burning liquid, for example, and spreads out quickly to form a blanket over it. This blanket cuts off oxygen to the fire, as well as cooling it down. The hose setups that supply water for fire suppression often have a small hose that can feed a foaming agent into the water stream.

The U.S. Air Force has developed a new class of powerful fire suppression agents known as encapsulated micron aerosol agents, or EMAAs. EMAAs are stored in the form of a solid, powder, or gel. When they are ignited, they form an *aerosol*, a suspension of very small particles in a gas. Smoke and fog are types of aerosols. The aerosol is lighter than air and smothers fires in enclosed areas. EMAAs are also effective against fuel tank fires. The Air Force has put them at the bottom of a fuel storage tank and then set them off when the fuel catches fire. The resulting aerosol percolates to the surface of the fuel and quickly extinguishes the fire. EMAAs are undergoing tests for environmental and occupational safety.

Removing the Oxygen

Methods of extinguishing fires remove at least one of the elements necessary for combustion (heat, fuel, or oxygen) or they interrupt the chain reaction of burning.

Breaking the Chain Reaction Some dry chemicals and hydrogenated hydrocarbon gases called halons extinguish a fire by interrupting the chemical reaction of the fuel with oxygen, which stops the flames. This method is effective on liquid and gaseous fuels because they must flame to burn. Although halons are very effective fire-extinguishing agents for certain types of fires, they are illegal in most drilling areas for environmental reasons. Only facilities that had halon systems already in place before the gas was outlawed can use it. Replacements are now available, but halons are still two to three times more effective.

Classifying Fires

Choosing which method to use to extinguish any particular fire relies on a system for classifying fires developed by the National Fire Protection Association (NFPA). The system classifies fires into four types, according to the type of flammable material.

- *Class-A fires:* Fires that involve common ash-producing materials, such as wood, paper, cloth, rubber, and some plastics (fig. 85). Water and foams are the best extinguishers for a Class-A fire.

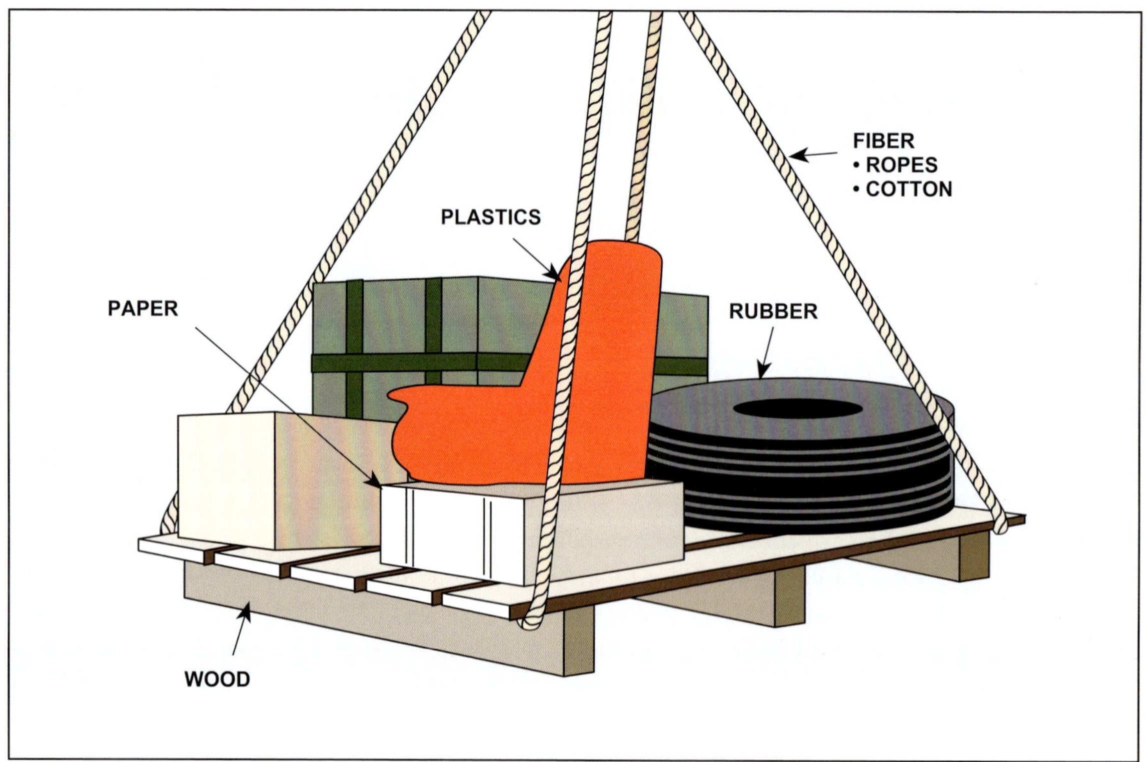

Figure 85. Class-A fires involve common flammable materials.

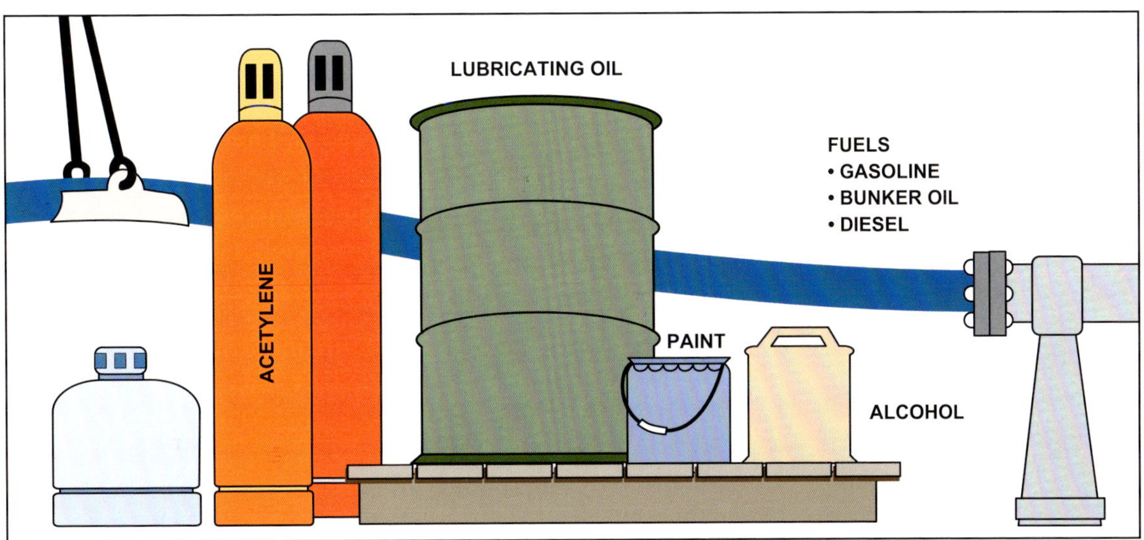

Figure 86. Class-B fires involve flammable liquids, gases, and petroleum products.

- *Class-B fires:* Fires that involve flammable liquids, gases, and greases, including petroleum and natural gas, as well as alcohols and paints (fig. 86). The best extinguishing method for Class-B fires is smothering them to cut off the supply of oxygen.

- *Class-C fires:* Fires that involve energized electrical equipment and lines (fig. 87). The extinguisher for these fires must *not* conduct electricity, so the best choice is to use CO_2 or a dry chemical rather than water to smother the fire or break the chain reaction.

Halons are an effective chemical means of extinguishing fuels, but they are illegal in most drilling areas.

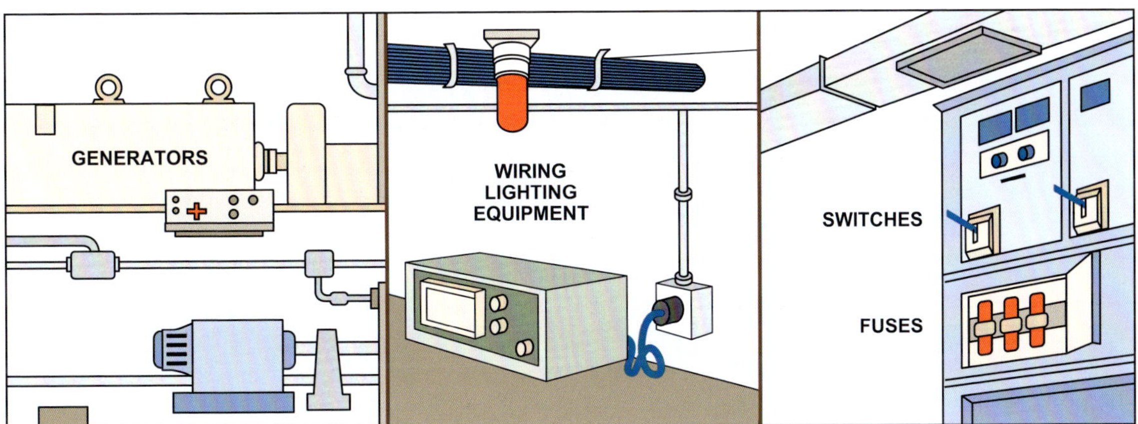

Figure 87. Class-C fires involve live electrical equipment and wiring.

There is an optimal method for extinguishing each of the four types of fire as classified by the National Fire Protection Agency.

- *Class-D fires:* Fires involving combustible metals, such as zinc, titanium, aluminum, and sodium. These metals are particularly hazardous when in powdered form, where they can become airborne and explode. They burn at such a high temperatures that water does not work as a cooling method (fig. 88). Special extinguishing agents are available for each metal to smother the fire. Fortunately, Class-D fires are rare on the drilling site.

In reality, some fires require the use of more than one extinguishing method. Firefighters extinguishing a Class-B fire that includes burning oil storage tanks and flow lines, for instance, use both water for cooling the metal and foam for smothering the flames.

Figure 88. Class-D fires can become airborne and explosive when in powdered form.

Fire suppression equipment on the drilling site includes portable and semiportable fire extinguishers and fixed systems for delivering water, foam, inert gases, or chemicals to the fire.

Fire Suppression Equipment

Portable extinguishers may be large enough to require a wheeled caddy (fig. 89) or small enough to carry (fig. 90). They are useful for quickly attacking a small fire. Even the larger extinguishers carry only a limited amount of extinguishing agent that is expelled very quickly, in 20 to 60 seconds.

Portable Fire Extinguishers

Figure 89.   A wheeled extinguisher has a hose for a longer range.

Figure 90.   A hand-held fire extinguisher has a short range of 6 to 8 feet (1.8 to 2.4 metres).

Labeling

Extinguishers are clearly marked as to the class of fire they will put out: A, B, or C (fig. 91). Some are rated for more than one class—AB, BC, or ABC. Do not use an extinguisher on a fire for which it is not rated.

If the extinguisher has an A or B classification, it also has a number marking that indicates its size or efficiency. The higher the number, the more fire the extinguisher can put out. For example, an extinguisher rated 4A will extinguish twice as much Class-A fire as one rated 2A. A 20B-rated extinguisher will extinguish five times as much Class-B fire as one rated 4B. Class-A extinguishers have number ratings from 1A to 40A; Class B from 1B to 640B.

A Class-C extinguisher has no number rating because this classification is simply added to a Class A or B rating to indicate that the extinguisher is safe to use on a fire involving electrical equipment.

A Class-D extinguisher is rated only for a particular combustible metal, and is never suitable for use on a Class A-, B-, or C-fire.

The Coast Guard uses a Roman numeral instead of an Arabic numeral to indicate the extinguisher's size. The numeral I is the smallest size, and V is the largest. For example, a BIII rating means that the extinguisher is medium sized and suitable for Class-B fires.

Figure 91. Fire extinguishers are labeled with a letter inside a geometric shape for their class (A, B, or C) as well as with a picture symbol for the type of fires they put out.

How to Use a Portable Extinguisher

The extinguisher's label should also have detailed directions on how to use it. In general, for a hand-held type, hold the extinguisher upright, pull the pin on top that keeps the handle from being pressed, aim the nozzle or short hose toward the fire, and squeeze the handle. Sweep the nozzle back and forth at the base of the flames.

Wheeled types work the same way—after wheeling the extinguisher to the fire, unravel the hose, and spray at the base of the flames.

Extinguishing Agents

The agent inside the extinguisher varies with the classification. Class-A extinguishers may contain water or a water solution, foam, or a dry chemical. Class-B extinguishers contain foam, a dry chemical, or CO_2. The agent in a Class-D extinguisher is a dry powder (not a dry chemical).

Many portable extinguishers use a cartridge (fig. 92) of nitrogen or carbon dioxide to supply pressure to expel the water, foam, or chemical inside. This cartridge is replaceable, and the extinguishing agent can be refilled.

Figure 92. A separate cartridge filled with pressurized nitrogen or carbon dioxide expels dry chemical extinguishing agent.

Inspection

To be sure that a portable extinguisher will be in working order when it is needed, make the following checks at least once per month:

- Check that the extinguisher is in the proper location and is accessible.

- Make sure the operating instructions and labeling are readable.

- Inspect the nozzle for obstructions, such as an insect nest.

- Look at the hose and fittings for corrosion or cracks.

- Check the lock pin to be sure that no one has tampered with the extinguisher.

- Make sure the extinguisher is full by weighing it or inspecting the level. Also check the pressure by looking at the pressure gauge. If it is not full and fully pressurized, replace it or recharge it.

- Check the inspection tag with your name and the date of the inspection.

Some drilling operators hire a fire protection company to make these inspections, but those are not the ones who will be burned if a fire gets out of control because of a faulty or damaged extinguisher.

Fixed Systems

Fixed fire suppression systems are permanently installed at certain locations on the drilling rig. The category includes several different types of delivery systems and extinguishing agents:

- A fire-main system for supplying water

- Foam systems

- CO_2 systems for flooding enclosed spaces

- Automatic sprinklers

Fixed systems are the big guns of fire suppression on a drill site.

Maintenance

Once a month, check the following:

- Make sure all the parts of the fire-main system—nozzles, hoses, portable monitors, and tools—are where they belong.

- Check the monitors for leaks, and that the valve handle moves freely between the open and closed positions.

- Make sure the swivels on monitors rotate while water is flowing.

- Lubricate mechanical components as recommended by manufacturers.

- Check that hoses are not kinked and that the threads are free so that the hose can be connected to the nozzle and hydrant by hand.

- Check the gaskets in the nozzles for cracks and wear, and clean out any obstructions inside the nozzle.

- Test the nozzle's bail and adjustment ring for free movement.

Each fire station also has a foam station, a tank of foam concentrate with special nozzles and a device that feeds foam concentrate into the water stream. In particular, the helipad will have a foam station as may processing, storage, or handling areas where foam is useful for controlling spills of flammable liquids. The foam system can be portable (carried from one station to another) as needed or it can be a fixed system in its own foam room with large containers of foam concentrate.

Once a month, check fire suppression equipment to make sure that the agents can still flow and there are no leaks.

Foam System

CO_2 System

A CO_2 system works by replacing the oxygen in the room. So personnel must evacuate the area where a system is activated in order to keep breathing.

Storage areas and rooms housing the engines, mud pumps, generators, compressors, and electronic equipment usually have a CO_2 system, which replaces the oxygen in the room with nonflammable CO_2. Before halons were banned, these areas may have had halon systems. Most rigs went back to using CO_2, but a few use other gases developed to replace halons.

The system consists of a group of cylinders full of pressurized carbon dioxide connected to a manifold and connected to each other with pressure hoses. The manifold has a master stop valve, which opens to let the CO_2 flow down a network of pipes toward the discharge nozzles located in the protected space. The number of cylinders, pipe size, number of nozzles, and their diameter depends on the size of the room and the concentration of carbon dioxide needed to protect it. The entire system operates using pressure, so no external power is necessary.

The CO_2 system may be manual or automatically activated by a fire detector. When activated, a warning alarm sounds, and the CO_2 first flows into a pneumatic timer cylinder. This timer delays the release of the gas for 20 to 40 seconds, giving any people in the room time to get out. (Cutting off oxygen to a fire cuts off oxygen for breathing at the same time.) The detector generally closes any exhaust fans and dampers automatically to seal the room so that the CO_2 cannot escape.

Automatic Sprinklers

The living, recreation, office, and galley spaces on a fixed offshore unit may be protected by an automatic sprinkler system. This system consists of piping, sprinkler heads, valves, pumps, and a pressure tank.

When fire causes one or more sprinkler heads to open, the initial water supply comes from the pressure tank. As the water flows out of the tank, the water itself or the pressure drop in the tank causes alarms to sound and starts the fire pump, which supplies more water to the system.

Sprinkler systems are reliable if they are well maintained. When one fails to operate, it is usually because the wrong valves are closed. To avoid this, mark the operating position for all valves and, if necessary, seal them in the proper position.

Some offshore rigs use a water spray system to protect the wellhead. The water spray system uses pumps, piping, and valves to send water through spray heads that provide an umbrella of water over the protected area. Other heads direct water to nearby walls or equipment to cool it. This system can be manual or automatic. When a detector activates this system automatically, the fire pumps begin pumping and a *deluge valve* opens. A deluge valve opens fast and allows the full volume of water to flow immediately.

Water Spray Systems

Since fire is a continuous threat on drilling rigs, most have automatic fire detection systems. The detector sounds an alarm when it is activated, and, in unattended areas, automatically turns on the fire extinguishing system. The rig also has a manual fire alarm system, and crewmembers can report fires by telephone and intercoms.

Fire Detection Equipment

Also called a *pneumatic tube fire detector*, this system is the only one that can detect fires in open spaces. It consists of a length of flexible plastic or metal tubing that forms a loop around the outside of whatever structure it is protecting. The tubing is filled with air or gas under pressure. The ends of the tubing are connected to pressure switches. When a fire burns through the tube, the pressure drop causes the pressure switches to flip on. The switches may control alarms and shutoff valves, and may turn off machinery and equipment and turn on fire-main pumps.

Fire Line Automatic System

Heat and Smoke Detectors

Heat and smoke detectors are spot detectors, like ordinary household smoke alarms. They are installed mainly in living spaces, rooms housing electronic equipment, and storage areas. These types of fire detectors are useless in open areas, where wind could carry away the heat and smoke of a fire. Usually they only sound an alarm when they sense fire.

Combustible-Gas Detectors

Another type of detector used in living areas, galleys, and equipment rooms is the combustible-gas detector (fig. 93). It may also protect product pipelines, manifolds, and the wellhead. This detector warns of dangerous concentrations of combustible or flammable gas. While it is not a fire detection system as such, it does detect situations that could lead to an explosion and fire.

One type of combustible-gas detector draws air into a chamber and heats it. Any combustible gases in the air sample burn, which changes the electrical output of a circuit in the unit. A meter shows the concentration of the combustible gas, and when the concentration reaches the danger point, the detector sets off an alarm.

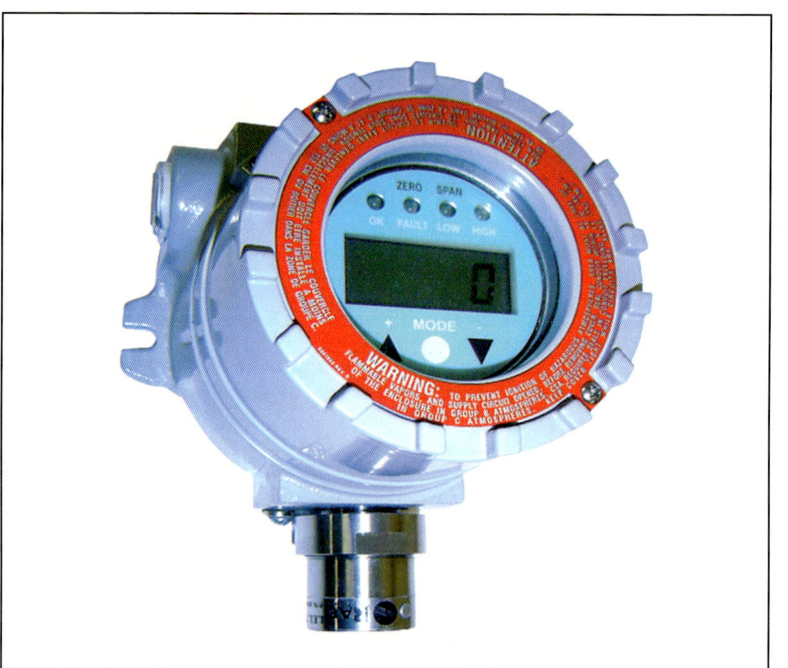

COURTESY OF RAE SYSTEMS

Figure 93.   A combustible-gas detector senses the presence of flammable vapors in the surrounding air.

No matter how many detectors a rig has, the crew should always watch for fires. Many times, an alert crewmember discovers a fire before even the most sophisticated detection system does. In spaces that have no such device, the crewmembers are the fire detectors. When someone sees a fire, he or she must go to the nearest manual alarm and sound it immediately. The most common manual alarm system is electrical—powered by the rig's generators, with batteries as a backup power source. The system includes fire alarm boxes (fig. 94) on all levels, in open and enclosed spaces, usually with telephones next to them.

Each alarm box has three buttons—yellow or orange, red, and black. Pushing the yellow or orange button sounds the fire alarm, a warbling two-pitched siren. This sound means that everyone should report to their fire stations. Pushing the red button produces a steady-pitched siren that is the signal to abandon the rig. The black button shuts off the siren; it is used if the alarm was set off accidentally or in error.

Manual Fire Alarms

Types of fire detection equipment:
- Fire line automatic systems
- Heat and smoke detectors
- Combustible-gas detectors
- Manual fire alarms

Figure 94. A rig has several manual alarm boxes.

Personal Safety Equipment

Personal safety equipment is a last defense for crewmembers against fire. Training emphasizes fire prevention.

The most important fire safety equipment any rig worker can have is thorough training in the prevention of fires, the location and operation of alarms and extinguishers, and when to use them. Those who man hoses and fixed fire suppression systems need additional training. Particularly on an offshore rig, there is no one but the drilling crew to put out a fire and nowhere to go to escape it except the ocean. Weekly fire evacuation drills take place at random times—usually, it seems, while sleeping after working a 12-hour shift.

On a rig, firefighters have access to protective clothing called a *bunker suit*. A rig usually has 6 to 12 protective suits. Additionally, all rooms have life jackets.

Bunker Suit

The bunker suit (fig. 95) consists of:

- Helmet—protects the head from impact and puncture injuries; the face shield protects the eyes from flying solids or liquids.

- Hood—protects from heat the parts of the face, ears, and neck not covered by the helmet or coat.

- Coat and pants or long coat—protects the body against cuts, abrasions, and burns, and provides limited protection against corrosive liquids.

- Gloves—protect the hands from cuts and burns.

- Boots—protect the feet from burns and puncture wounds.

- Air supply—provides oxygen to breathe and protects the lungs and respiratory tract from dangerous gases, heat, and smoke.

Coat and Pants

The coat and pants are made of three layers—an outer shell, a moisture barrier, and a thermal barrier. These layers protect against flame, cold temperatures, and hot water and vapors. If the wearer does not suit up properly, however, he or she will not be completely protected. For example, the coat has a double-layered storm flap closure in front. Be sure to close both the snaps or zipper inside and the Velcro flap outside. Just closing the flap and not the inner closure could allow the coat to fly open. Closing only the inner closure and not the flap allows water and steam to get inside between gaps.

Figure 95.   A bunker suit reflects as much as 90% of the radiant heat from a fire.

Air Supply

Because fires consume oxygen, an air supply is crucial to the firefighter. Fires also release carbon monoxide and small particles of carbon, tar, and dust that form smoke. Inhaling carbon monoxide or smoke can quickly disable a person. The air supply (fig. 96) also protects the wearer in a room with a CO_2 system that has been discharged, and in areas where burning materials produce dangerous gases such as hydrogen cyanide and hydrogen chloride.

The air supply consists of a cylinder of compressed air or oxygen, straps to hold it onto the wearer's back, a face mask with hose, and a regulator assembly to regulate air pressure. Cylinders come in different sizes; a typical one is rated for 30 minutes of air and weighs from 9.6 to 23.8 pounds (4.3 to 10.7 kilograms), depending on whether it is aluminum or steel. The regulator reduces the pressure of the air from the cylinder to slightly above atmospheric pressure and controls the flow of air. Inhaling moves a diaphragm out so that air can move in, and exhaling moves the diaphragm back to the closed position. A pressure gauge that shows the pressure of the air remaining in the cylinder

Figure 96. Air supply equipment

is mounted next to the regulator. An alarm sounds when the pressure falls below a certain amount. This means that the air supply is about to give out, and the firefighter should leave the area of the fire immediately.

The face piece is made of clear plastic with a rubber mask that fits snugly against the face by means of straps. The hose brings air from the regulator to the face piece, so be sure to keep it free of kinks. The hose is corrugated to keep it from collapsing when the wearer leans against a hard surface.

A label on the coat will give instructions on how to keep it in good shape.

Maintenance of Protective Clothing

- Wash all oil, grease, chemicals, and dirt from the bunker suit, boots, and face piece after use. Contaminants may cause the suit fabric to deteriorate, and dirt absorbs heat faster than the protective outer layer of the suit.

- Repair tears according to the manufacturer's instructions.

- Store the clothing properly.

- Make sure the air cylinder is full.

- Check all gauges and the low-pressure alarm on the air supply.

- Check the face piece and hose by inhaling with your hand covering the end of the hose.

- Connect the hose to the regulator and inhale to check the regulator.

- Check that all straps are in good condition.

To summarize—

Fire-prevention measures on the rig

- Avoid exposing flammable materials to sources of ignition; take special precautions when smoking.

- Investigate unusual odors and smoke.

Science of fire

- A fire is ignited when a vapor is exposed to heat and oxygen, starting a chemical reaction of oxidation.

- There are several ways to extinguish fires: remove the fuel, remove the heat, remove the oxygen, or interrupt the chain reaction of the fuel with the oxygen.

Fire suppression equipment

- Fire extinguishers are labeled for their ability to fight different classes of fires (A, B, C, and D) and can be hand-held or wheeled.

- Foam systems blanket the fire, removing oxygen.

- Carbon dioxide systems remove the oxygen from the air in an enclosed space.

- Fire-main, sprinkler, and water spray systems flood the fire with water. Fair-main and sprinkler systems are fixed in place; water spray systems are set up as stations where human operators direct the water spray.

Fire detectors

- Fire line automatic systems use pressurized tubing that can encircle an open space.

- Heat and smoke detectors sense high temperatures and the presence of smoke in an enclosed space.

- Combustible-gas detectors warn of a high concentration of any flammable gas.

- Manual fire alarms stationed around the rig are used when a crew member detects a fire.

Personal protective equipment

- A bunker suit protects the wearer from burns, smoke, dangerous gases, and flying objects and liquids.

- A bunker suit includes a helmet, hood, coat and pants or long coat, gloves, boots, and an air supply.

Hydrogen Sulfide Safety

In this chapter:

- Areas of the rig where hydrogen sulfide gas is encountered
- Hazards of H_2S gas
- The need to use equipment to detect H_2S
- Preventive procedures and personal protective equipment
- Emergency evacuation and first aid for gas exposure

Hydrogen sulfide is a deadly gas. Also called sour gas or sulfur gas, it is common in some areas of the oilfield. It is more deadly than carbon monoxide and almost as toxic as hydrogen cyanide. When a well is drilled into a formation containing H_2S, the gas may be circulated to the surface in the drilling or formation fluids. Several areas around the rig then become potentially dangerous. One such area is around the bell nipple. The entire circulation system—shale shakers, pits, tanks, mud lines, pumps, and other components—has a high potential for H_2S contamination. Leaks in hoses, lines, and connections can create hazardous, toxic conditions.

The deadly gas H_2S can be hard to detect without instruments because it is colorless and, at a certain concentration, deadens the sense of smell.

H_2S gas is heavier than air (having a density of 1.189 compared to 1.0 for air) so it tends to accumulate in low areas around the rig such as the cellar, ditches, and open mud troughs. Low areas are especially hazardous when there is no wind or fans to disperse the gas.

H_2S is dangerous because it is colorless and can burn or explode once in the air. Restrictions on smoking, burning, or welding must be strictly observed. It burns with a blue flame that produces another dangerous gas, sulfur dioxide, or SO_2. In low concentrations, H_2S has the odor of rotten eggs. Smell cannot be relied on, however, to detect H_2S because the gas quickly destroys the sense of smell.

Characteristics of H_2S Gas

H$_2$S is soluble in liquid. This is extremely important because it means concentrations of the gas may be contained in pools of water, in containers, or in sludge at the bottom of a tank, where it can be released when the pool is agitated or heated.

The gas is highly corrosive to steel. Further, H$_2$S reacts with iron in steel to form iron sulfide. Iron sulfide reacts with hydrochloric acid to form H$_2$S gas. Precautions should therefore be taken during acid jobs, even in old wells where the gas has not been found before. In fact, recent drilling in old fields and waterflood areas has experienced an increase in H$_2$S incidents because of the formation of iron oxide.

Hydrogen sulfide gas can only be detected by smell at very low concentrations. The following figures give an idea of how small the window for detection is. The unit *ppm* signifies *parts per million* by volume of gas to air; 1 ppm of gas, or 1 part gas in 1 million parts air and gas, would compare to 1 inch in 15.5 miles.

1 ppm	*Minimal perceptible odor*
5 ppm	*Easily detectable, moderate odor*
10 ppm	*Eye irritation*
27 ppm	*A strong, unpleasant odor, but not totally intolerable*
100 ppm	*Coughing, eye irritation, loss of sense of smell, stinging eyes and throat*

How much hydrogen sulfide can be inhaled before it causes damage? The effects of H$_2$S depend on the length of exposure and the concentration of the gas. Table 2 shows how the toxic effect multiplies rapidly with increasing concentration until instant unconsciousness and death occur at 1,000 ppm (equivalent to only $\frac{1}{10}$ of 1%).

How does H$_2$S work in the body? When a person breathes in H$_2$S, it goes directly through the lungs and into the bloodstream. When the amount of gas breathed in exceeds the amount the body can break down, the H$_2$S builds up in the blood, and the nerve centers in the brain that control breathing are paralyzed; the lungs stop working, and the person is asphyxiated.

An individual may be hypersensitive to H$_2$S because of a health problem; therefore, advise the rig manager of any health problems you may have before any potential exposure to H$_2$S. A perforated eardrum precludes anyone from using a breathing apparatus. Among

Table 2
Toxicity of Hydrogen Sulfide Gas

Concentration (ppm)	Concentration (percentage)	Toxic Effects
10	$\frac{1}{1{,}000}$ of 1%	Does no damage for up to 8 hours' exposure
100	$\frac{1}{100}$ of 1%	Destroys sense of smell in 3 to 15 minutes May sting eyes and throat
200	$\frac{2}{100}$ of 1%	Kills sense of smell quickly Stings eyes and throat
500	$\frac{5}{100}$ of 1%	Causes loss of reasoning and balance Produces respiratory paralysis in 30 to 45 minutes Demands prompt artificial resuscitation
700	$\frac{7}{100}$ of 1%	Checks breathing, with consequent death unless rescue is prompt Requires immediate artificial resuscitation
1,000	$\frac{1}{10}$ of 1%	Causes instant unconsciousness *Makes permanent brain damage highly probable without immediate rescue*

the conditions that may compromise a person's tolerance to H_2S are pulmonary, respiratory, bronchial, and heart problems. Other medical considerations include an eye infection, diabetes, epilepsy, hypertension, and alcoholism (or consumption of alcohol within 24 hours of exposure). Persons with any of these conditions should not be placed in jobs involving exposure to H_2S or SO_2.

Precautions Against H₂S

Figure 97.   Worker using hand-held H₂S detector

There are ways to protect the crew from the dangers of H_2S. Drilling mud should be heavy enough to contain any H_2S gas kick. An H_2S neutralizer added to the mud can help prevent the gas from reaching the surface. Detectors can be placed at the shale shaker, bell nipple, tanks, or other places where H_2S is likely to be first noticed. The detectors will trigger an alarm system. Hand-held detectors should also be available and used periodically to check the atmosphere around the rig (fig. 97).

The crew should know the location of all breathing equipment. It should be checked periodically, and personnel should receive training in its proper use.

Wind easily disperses H_2S, and windsocks, flags, or streamers indicate the wind direction. Blowers and fans should be in place for use when the wind is calm.

A designated safe area should be established where the crew can assemble for instructions during an H_2S emergency. Designating two areas ensures that there is one safe area no matter the wind direction. Set up a buddy system to assure that all persons will be accounted for in an emergency.

A sign and flag system should be in place at the entrance to the location to warn of H_2S conditions on the drill site. Warning flags should be:

Green	Normal operations
Yellow	Moderate amounts of H_2S may be present. This flag is put up when drilling reaches 1,000 feet (305 metres) above the potential hazardous zone.
Red	Extremely hazardous conditions; H_2S is present. The red flag is flown regardless of the concentration if gas or an odor has reached the surface.

Emergency telephone numbers or radio contact procedures should be posted so medical personnel, company management, and authorities can be notified quickly when necessary.

Crews should receive training in order to learn the procedures to follow in the event of an H_2S encounter. A well-prepared crew will know what to do when H_2S is encountered and is less likely to panic in an emergency situation.

When gas is encountered, move away from the rig while facing into the wind to quickly escape the danger area. H_2S accumulates in low areas, so also move in an uphill direction, if possible. Never move toward the wellhead, the source of the hazard, even when it is upwind. Assemble at the previously established safe area and await instructions.

If there should be a sudden release of H_2S, the crew should not panic. They should hold their breath, put on breathing equipment, help anyone in distress, and go to a designated, upwind safe area.

If persons have been overcome, they need immediate rescue (fig. 98), but do not attempt a rescue procedure until the rescuer has put on protective breathing equipment. Quickly move the victim to a place where the air is safe to breathe. Notify a supervisor immediately so that arrangements may be made to transport the victim to a hospital.

Emergency Procedures

When H_2S is detected, the crew should quickly put on breathing equipment, help anyone in distress, and evacuate the area.

Figure 98. Do not attempt a rescue without protective breathing equipment.

Breathing Equipment

Figure 99. Rescue breathing unit

There are many brands and types of protective breathing equipment available; regardless, all supplied air respirators must function in a positive-pressure mode. Never use air-purifying respirators in an H_2S situation. Breathing units can be divided into three categories:

- Escape units have a self-contained air supply (grade D) that gives a person enough time (about five minutes) to get to a safe area in an emergency.
- Rescue units have a self-contained supply of air in bottles that are usually worn on the back (fig. 99). The average air supply will last about 30 minutes but varies according to the wearer's exertion and breathing pattern.
- Work units are usually supplied with air from a central supply that allows the wearer to work safely in an H_2S environment.

Before putting on protective breathing equipment, personnel should take certain precautions. Health problems, as previously discussed, should be considered. Positive-pressure air masks are individually fit-tested to assure an airtight fit. Several things can affect the fit. Denture wearers should be fitted while wearing the dentures for a proper fit; they must then wear the dentures when using the unit. Anyone using a breathing unit should not wear a beard, mustache, long hair, eyeglasses, or contact lenses. These things may interfere with the fit and efficiency of the face mask.

To summarize—

Precautions to avoid hydrogen sulfide poisoning

- Place detectors near equipment likely to be contaminated.
- Put breathing equipment, warning signs, fans, and windsocks or streamers in the appropriate places.
- Designate a safe area for the crew to assemble.

Emergency procedures for H_2S exposure

- Move away from the wellbore facing the wind.
- Put on breathing equipment.
- Help anyone in distress.

First Aid

First aid is the immediate, on-site care given to a person who has been injured or becomes suddenly ill. It includes self-help when medical assistance is not readily available or is delayed. Drilling operations are frequently in remote locations, so professional medical care may be hours away. Knowledge of basic first aid and cardiopulmonary resuscitation (CPR) is critical because prompt and correct care can mean the difference between life and death or between rapid recovery and long hospitalization. Even minor delays in caring for an injury or sudden illness can be fatal. To ensure prompt care, the rig manager should contact local medical facilities and evacuation services soon after moving on location. Information for emergencies should then be posted in the doghouse and other designated places on the rig. All accidents, even minor ones, should be reported to the rig manager. The crew should understand that, except for very minor injuries, first aid should be followed by treatment at a medical facility.

Ideally, every employee would be trained in first aid and know what to do—and what not to do—in a medical emergency (fig. 100). With the frequent changes in drilling personnel, it may not be practical to enroll every employee in a first aid course, but there should be one person knowledgeable in first aid and CPR on each tour.

Emergency First Aid

There should be at least one person trained in first aid and CPR on each tour.

Avoid moving a seriously injured victim if he or she can safely remain at the scene of the accident and await treatment.

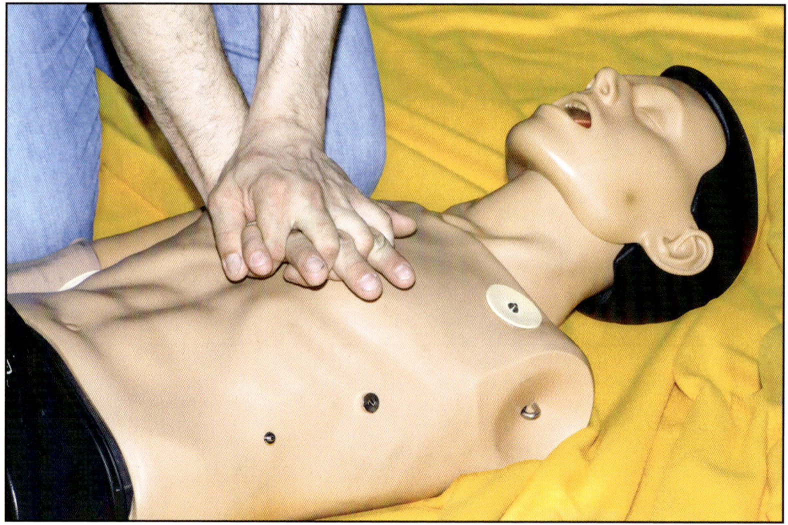

Figure 100. Ideally, all crewmembers are trained in CPR.

Some injuries are life threatening. They require immediate care in cases of severe bleeding, heart failure, stoppage of breathing, or poisoning. In less serious situations, attention should be given to preventing further injury and getting assistance. A caregiver's calm, reassuring manner is important to the victim.

There are three basic steps to take in any emergency:

1. Check the scene and the victim.
2. Call 911 or the posted emergency number.
3. Care for the victim.

CHECK—Before you can help the victim, you must make sure the scene is safe for others and yourself.

1. Look the scene over and determine:
    A. Is it safe? Are there fires, spilled chemicals, poisonous gases, electrical lines, collapsing structures, or smoke? If these, or other dangers, are present do not go near the victim. Stay a safe distance away and call the local emergency number immediately. Leave dangerous situations to the professionals.
    B. What happened? Look around for clues. Is there a fallen ladder or an electrical line on the ground? Knowing what happened will influence the care to be given.
    C. How many victims are there? If there are several, who is in greatest peril? An unconscious victim is in greater danger than one who is screaming.

    D. Can others help? Did anyone see what happened? Can someone cut off running machinery and fuel lines or call for medical assistance?

2. Check the victim(s):
   A. Conscious
   B. Unconscious

CALL—Use 911 or the posted emergency number. This is often the most important action you can take to help the victim. The caller should know the victim's condition to relay that information to the dispatcher. Do not hang up before the dispatcher does to make sure all the needed information has been given.

CARE—In cases of serious injury or sudden illness, quickly determine the best way to rescue the victim without risking further injury. Movement is a dangerous threat to the seriously injured and should be avoided unless there is imminent danger from fire, explosion, collapse, or other peril. Treatment should include the following procedure:

1. Treat for unconsciousness. Keep an air passage open, keep the victim breathing, and maintain/restore a pulse.
2. Control severe bleeding.
3. Treat for shock.
4. Immobilize fractures.

Supplies and Equipment

Every rig should have a copy of the American Red Cross manual "Adult First Aid/CPR/AED" (or an equivalent) and one or more first aid kits or cabinets (fig. 101). Personnel should know the location of each of these items and be familiar with them. Sealed, sterile dressings should be discarded and replaced if the wrapping has been broken.

At least one stretcher should be available on location. Larger rigs and offshore units, where the floor is considerably above the ground or water surface, should have litters or baskets that allow a victim to be lowered to the ground while maintaining the immobility needed to prevent further harm.

Splints are needed to immobilize fractures and should be available on site. Two or more blankets should be available for use in treating shock. Other equipment may be needed to meet special conditions.

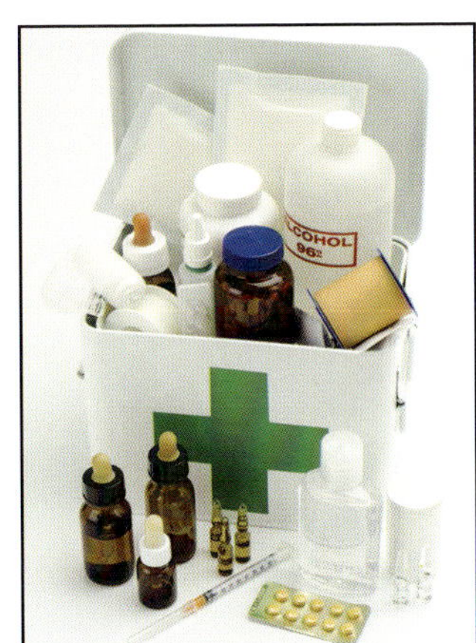

Figure 101. Every rig should have a first aid kit.

147

Training First aid training, including CPR, gives employees the knowledge and skills to give life support and other kinds of emergency care. Crew training can save lives.

Each crewmember should know how to rescue anyone overcome by toxic gas or contaminated air. Be aware, however, that the rescuer must wear respiratory equipment and a lifeline before entering the hazardous area. (See the chapter on confined spaces, page 93.)

Treatment in the Field The following brief discussion on common medical emergencies is not meant to substitute for formal first aid training or to supplant the American Red Cross manual "Adult First Aid/CPR/AED," which should be found on every rig.

Unconscious Victims Check the victim for:

- Breathing
- Pulse (fig. 102)
- Bleeding

Call or have someone call for medical assistance; then initiate rescue breathing (formerly termed artificial resuscitation) at once. Time is critical. You may not know when breathing stopped; brain damage is possible in 4 minutes, and severe, irreversible brain damage or death in 10 minutes.

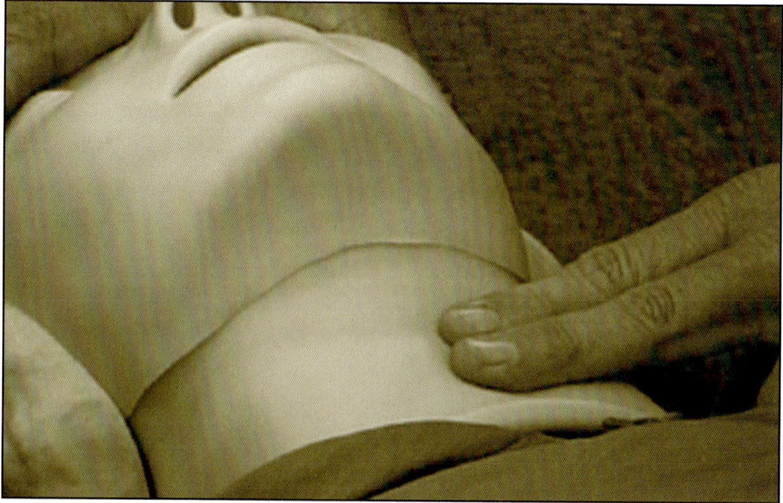

Figure 102. In an unconscious victim, check for pulse.

Profuse bleeding must be stopped immediately. If a large blood vessel is cut, death may occur within a minute or two. Do not waste time washing the wound. Control severe bleeding by placing a sterile dressing or a clean cloth over the wound and apply direct pressure (fig. 103). Any cloth will do, but the cleaner the better.

Bleeding

Bleeding and shock are serious conditions. Seek medical help immediately and treat the victim at the scene.

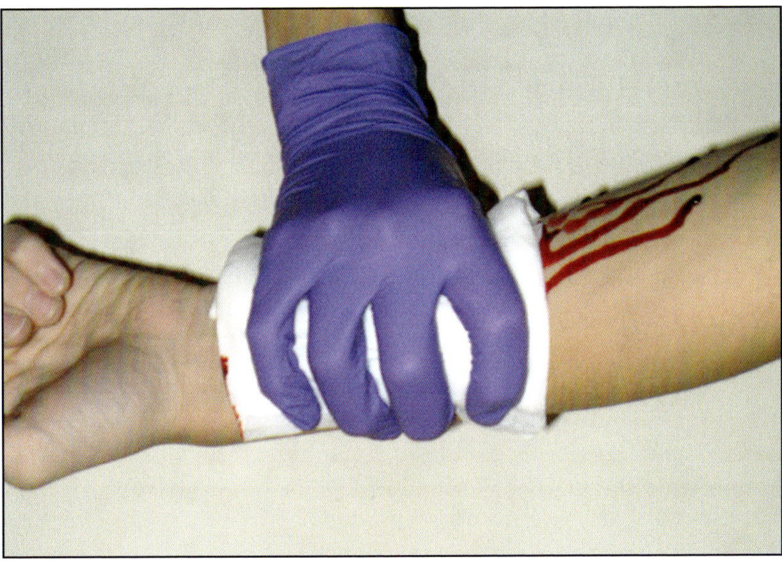

Figure 103.   Control bleeding with direct pressure.

Severe bleeding, serious injury, temperature change, or significant loss of body fluids can lead to a life-threatening condition called shock. Shock is a condition in which the circulatory system fails to deliver blood to all parts of the body. This triggers a domino effect: Organs fail to function from lack of blood and oxygen, arm and leg tissues die, blood is directed away from the brain, heart beat becomes chaotic, the heart fails to pump blood and stops, breathing stops, and death quickly follows.

Shock

Symptoms

You cannot always determine the cause of shock but you can recognize the symptoms:

- Restlessness or irritability is often the first indicator that an injured person may be going into shock. He or she may complain of severe thirst and vomit or retch from nausea.

- There is an altered consciousness or awareness.

- The victim's skin is cool, moist, and pale or bluish.

- Changes in the breathing pattern appear. Breathing may become rapid and shallow or deep and irregular.

- The pulse becomes rapid and faint, possibly too faint to be felt at the wrist but perceptible at the carotid artery at the side of the neck.

- In later stages of shock, the victim becomes apathetic and unresponsive. Eyes appear sunken and vacant, and the pupils dilate widely.

In some cases, the victim loses consciousness and his or her body temperature drops. These symptoms can be followed by death.

Treatment

Call for emergency medical help immediately. Shock cannot be managed by first aid alone. Advanced medical care is required as soon as possible. While awaiting medical help, the following care is recommended:

- Have the victim lie down. Being comfortable is important because stress or pain accelerates the shock progression.

- Control any external bleeding.

- Keep the victim warm and reassured.

- Elevate the legs about 12 inches (305 millimetres) if there are no head, neck, or back injuries or possible broken bones in the hips or legs. If in doubt, leave the person lying flat.

- Do not give the victim anything to eat or drink even though he or she may complain of thirst.

Burns are often caused by careless or inattentive actions with or around flammable agents. (See the chapter on fire detection and suppression, page 115). The most common burns are caused by sources of heat or flames. Electrical current can burn the body both internally and externally. Certain chemicals, like caustic or acid, can also cause burns.

Burns are described by their cause—heat, electricity, chemical, or radiation—or by their depth. The deeper the burn, the more severe it is.

Burns

Burns are described by their depth—first, second, or third degree.

First-Degree Burns

First-degree burns are superficial, involving only the top layer of skin. The most common causes are sunburn, contact with hot objects, steam, or hot water. The skin is reddish, dry, possibly swollen, and painful. These burns heal in 5 to 6 days and leave no permanent scar.

Second-Degree Burns

Second-degree burns are deeper than first-degree burns, involving the second layer of skin. The burned skin is red and blotchy. There are blisters that may open and weep clear fluid. The burn is painful and often swollen. The most common causes are severe sunburn, hot liquids, steam, and flammable fluids (which cause flash burns). Healing may take 3 to 4 weeks, and scarring may result.

Third-Degree Burns

Third-degree burns are critical burns. All skin layers and some or all of the underlying flesh, bone, and nerves are destroyed. The burn may appear brown or black, but underlying tissue sometimes appears white. These burns can be either extremely painful or fairly painless if the nerve endings have been destroyed.

Flames, burning clothing, hot liquids, hot objects, and electricity are common causes of third-degree burns. Areas of less severe injury, which may actually be the major source of pain, usually surround third-degree burns.

Treatment of Burns

The care of burn victims involves three basic steps:

1. STOP the burn by putting out the flames or by moving the victim away from the source of heat.

2. COOL the burn. Use large amounts of cool water. Immerse the burned area in cool water if possible; use a shower, tub, water hose, or whatever is available. Do not use ice or ice water except on small superficial burns. Apply soaked towels or sheets to areas that cannot be immersed in water.

3. COVER the burn. Use dry sterile dressings or a clean cloth secured loosely in place. Covering the burn helps keep out air, reduces pain, and lessens the danger of infection. If the burn covers a large part of the body, cover it with a clean, dry sheet or other cloth.

Refer to figure 104 for important DOs and DON'Ts of burn care.

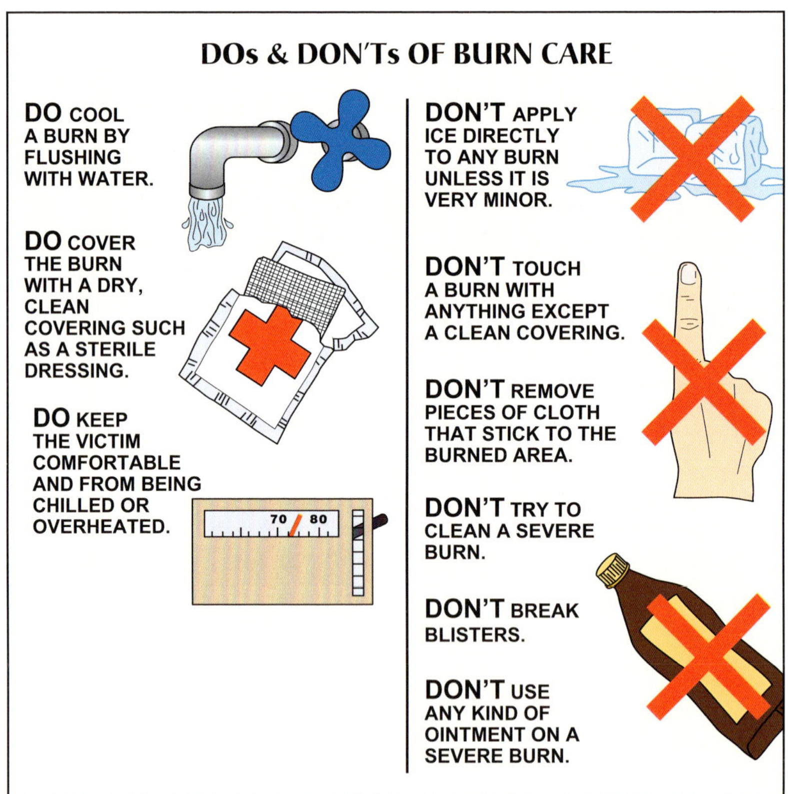

Figure 104.   The dos and don'ts of burn care

Do not break blisters or attempt to remove charred particles of clothing from critically burned areas. Do not attempt to clean a severe burn. If the hands are burned, try to keep them above the victim's heart level. Keep burned feet or legs elevated and do not allow the victim to walk. Do not use ointments, grease, or home remedies on severe burns. Such substances may cause complications or interfere with the physician's treatment.

Keep the victim comfortable and from being chilled or overheated. Be alert for signs of shock.

Chemical Burns

For chemical burns to the skin or eyes, wash away the chemical with water from a shower, hose, or eyewash fountain for at least 5 minutes or until emergency help arrives. Remove any clothing from the burn area and apply bandages. When moisture contacts caustic soda, a chemical reaction begins that can cause a burn upon contact with the body. Household vinegar can neutralize caustic soda on the skin, but do not use it for burns to the eye.

Electrical Burns

The severity of electrical burns depends on the length of the contact, the strength and type of the current, and the path the current takes through the body. There are often two wounds, one where the current enters the body and one where it exits. Electrical burns are often deep and appear minor but the tissue below the surface may be severely damaged. The burn itself is not likely to be the major problem. The victim may be unconscious and should be checked for breathing and pulse. Check for other injuries and do not move him or her.

Do not cool an electrical burn; instead, cover it with a dry sterile dressing. Keep the victim from getting chilled.

Take all electrical burns seriously, even when the victim does not appear to have serious burns on the outside.

Heat-Related Illnesses

Heat-related illnesses include, in order of severity, cramps, heat exhaustion, and heat stroke. Once the signals of a heat-related illness appear, the victim's condition can quickly become life threatening or even fatal. Increased salt intake during exposure to extreme heat can lower the incidence of heat-related problems, but do not take salt while experiencing a heat-related illness.

Heat cramps, while not dangerous, are often the first signals that the body is having trouble with the heat. Heat cramps are painful muscle spasms that usually occur in the legs or abdomen. Move the victim of heat cramps to a cooler place. Rest and fluids—either water or a sports drink—are all that are needed to recover. Do not give the victim salt tablets or salt water; they can make the situation worse.

Heat exhaustion is more serious than heat cramps. Symptoms of heat exhaustion include a subnormal temperature and cool, moist, pale, or flushed skin. The victim may complain of headache, nausea, dizziness, weakness, and exhaustion.

Heat stroke is the least common but most dangerous heat emergency. Heat stroke often occurs when the symptoms of heat exhaustion are ignored. The body systems are overcome by heat and stop functioning, which results in a serious, possibly fatal, medical emergency. The symptoms of heat stroke are the opposite of heat exhaustion and include an elevated temperature and red, dry, and hot skin. He or she may experience changes in consciousness, a rapid, weak pulse, and fast, shallow breathing.

Treatment

A heat-related illness is usually reversible if treated early. Treatment for heat exhaustion and heat stroke include these steps:

1. Get the victim out of the heat.
2. Loosen tight clothing.
3. Remove perspiration-soaked clothing.
4. Apply cool, wet cloths to the skin.
5. Fan the victim.
6. If conscious, give the victim cool water but do not let him or her drink too quickly: Give about 4 ounces every 15 minutes.
7. If the victim refuses water, vomits, or undergoes changes in breathing or consciousness, his or her condition is worsening. Call for an ambulance immediately. If there is vomiting, stop giving fluids and turn the victim on his or her side.

Continue treatment by keeping the victim lying down and cool. Ice packs may be placed on the wrists, ankles, arm pits, or groin to cool the large blood vessels. Do not apply rubbing alcohol.

Cold-Related Illnesses

Hypothermia and frostbite are two types of cold emergencies that result from prolonged exposure to cold. A person can develop hypothermia when the temperature is above freezing if strong winds and snow drop the chill factor dramatically. Anyone remaining in cold water or wet clothing for a long time risks getting hypothermia. In hypothermia, the whole body cools because its ability to keep warm fails. The victim may die without proper care. Frostbite is the freezing of body parts exposed to the cold. The severity of the damage depends on how chilling the wind is in addition to the length of exposure and the temperature. Frostbite can cause the loss of fingers and toes or even hands, feet, or legs. The victim of frostbite needs professional medical treatment as soon as possible.

Frostbite

Superficial frostbite commonly affects the ears, nose, chin, cheeks, fingers, and toes (fig. 105). Symptoms of frostbite include a lack of feeling and discoloration (white, blue, yellow, or flushed) in the affected area. The skin appears waxy and cold to the touch. Severe cases of frostbite can involve larger areas of the body such as hands, feet, or legs. In deep frostbite cases, the tissue is frozen deep below the surface and the area is white, waxy, and hard to the touch.

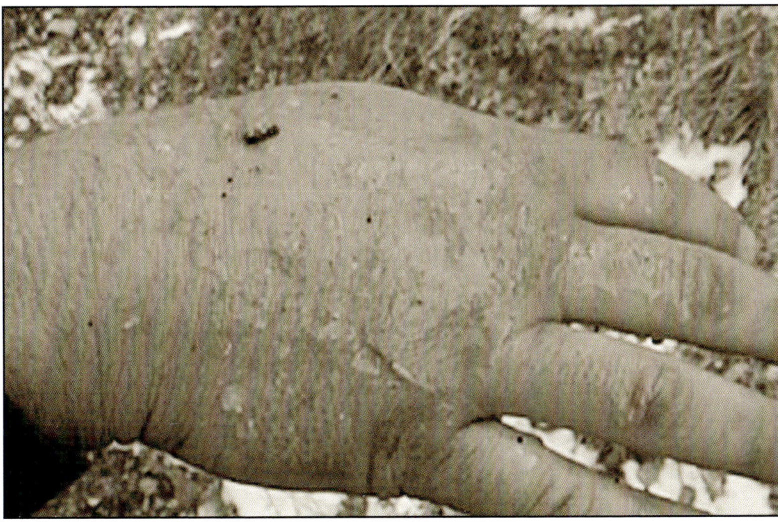

Figure 105. Superficial frostbite commonly affects the hands.

A victim of frostbite or hypothermia needs professional medical attention as soon as possible. In the meantime, remove the person from the cold and gradually warm him or her.

Care for frostbite requires moving the person from the cold and gradually warming him or her. Handle the affected area carefully. Never rub the frostbitten area. Rubbing will cause further injury to the soft tissues. Warm the affected part gradually by soaking it in warm water—water no warmer than 105°F (40°C). If the water is uncomfortable to your touch, it is too warm. Keep the part in the water until it looks red and feels warm. Do not break any blisters. Loosely bandage the area with a dry, sterile bandage. If fingers or toes are frostbitten, place cotton or gauze between them and secure medical attention as soon as possible.

Hypothermia

Signs of hypothermia include shivering, numbness, apathy, a glassy stare, and unconsciousness. To care for hypothermia, start by calling for medical help, then move the victim to a warm place. Handle him or her gently. Remove any wet clothing. Gradually warm the victim by putting him or her in dry clothing or wrapping in blankets. You may apply a heat source but keep a layer of clothing, towel, or blanket between the source and the body to avoid burning the skin. If he or she is conscious, give warm liquids. Do not warm the victim too fast by immersing him or her in warm water; this could cause heart problems.

In severe cases of hypothermia, the victim may be unconscious and the body may be stiff. Check for breathing and pulse and give rescue breathing or CPR, if necessary, until the medical emergency team arrives.

A foreign object lodged under the upper eyelid may be removed by pulling the upper eyelid down over the lower lid and releasing. Repeat as needed. If the object remains in place, flushing the eye with water may remove it. Do not attempt to remove an object lodged in the eyeball. Your effort may only force the object deeper and cause more serious injury. Place a compress over both eyes and seek medical attention at once.

Eye Emergencies

For chemicals or H_2S in the eye, flush the eye with water for 5 to 15 minutes (fig. 106). Force the eyelids open if necessary to get a complete flushing action. Prompt medical attention is recommended.

Figure 106. Special eye wash facilities

Heart Attack

When someone is having symptoms of a heart attack, call for medical assistance immediately.

Symptoms

The primary sign of heart attack is discomfort or pain in the chest that does not go away. The pain is most often felt in the center of the chest but may spread to the shoulder, arm, neck, or jaw. The symptoms may vary from discomfort to a crushing sensation in the chest. The victim may complain of pressure, squeezing, heaviness, or tightness in the chest. The pain is constant and is not relieved by rest, changing position, or medication. If pain lasts for more than 10 minutes, the victim needs medical attention at once. A brief, sharp, stabbing pain that gets worse with deep breathing or bending over, and subsides with rest, is probably not heart related.

Another signal of heart attack is difficulty in breathing and shortness of breath even though breathing may be faster than normal. The face may be pale or bluish and damp with sweat. Some victims sweat profusely. There may be changes in the pulse rate; it may be slower, faster, or irregular.

Care

A heart attack, once apparent, requires prompt action. Most people dying from heart attack do so within 2 hours of the first symptoms. Do not delay! Call for medical assistance immediately.

Encourage the victim to rest. A sitting position is usually best, especially if there is difficulty in breathing. Be calm and comforting and try to obtain information about his or her condition. Closely observe the victim and watch for any changes in behavior. The victim can go into cardiac arrest at any time; when that happens, breathing stops and there is no pulse. CPR, a combination of rescue breathing and chest compressions, must be started at once and continued until medical help arrives.

Fractures of the bone may be simple or compound. In simple fractures, the bone remains in line and does not puncture the skin; it may be merely cracked. In a compound fracture, the bone is completely broken and may protrude from the skin.

There are several indications of a fractured bone. The victim may feel or hear the bone snap or grate when moved, and movement may be difficult and painful. Swelling, discoloration, deformity, or changes in shape or length of bones may be observed.

Care

The victim should be checked for signs of shock and treated accordingly. Protect him or her from further injury by immobilizing the injured part. The fractured arm or leg should be elevated if it can be done without disturbing the fracture. Do not attempt to set the fracture. If the bone protrudes through the skin, do not attempt to push the bone back or replace bone fragments. Do not wash or probe the wound but cover it with a sterile bandage, towel, or other clean cloth.

Splints are devices used to immobilize fractured bones. Apply splints only if the victim must be moved or transported by someone other than emergency medical personnel. Do not splint a fracture if doing so causes greater pain and discomfort. Commercial splints are available but they may be fashioned from any stiff material. The splint should be long enough to extend past the joints on either side of a suspected fracture. Padding on the splint will improve the victim's comfort.

Fractures

If a fracture or a break is suspected, check the victim for signs of shock. Treat for shock, if necessary, and immobilize the injured body part as soon as possible with a splint.

One way to immobilize an injury is to use the victim's body as a splint (an anatomic splint). For example, an uninjured leg can serve as a splint for a fractured leg (fig. 107); the chest can serve as splint for a fractured arm.

Do not tie the injured part too tightly to the splint. Swelling or discoloration of the fingers or toes indicates the splint is too tight. If the victim cannot feel his toes or fingers or feels a tingling sensation, the ties should be loosened immediately.

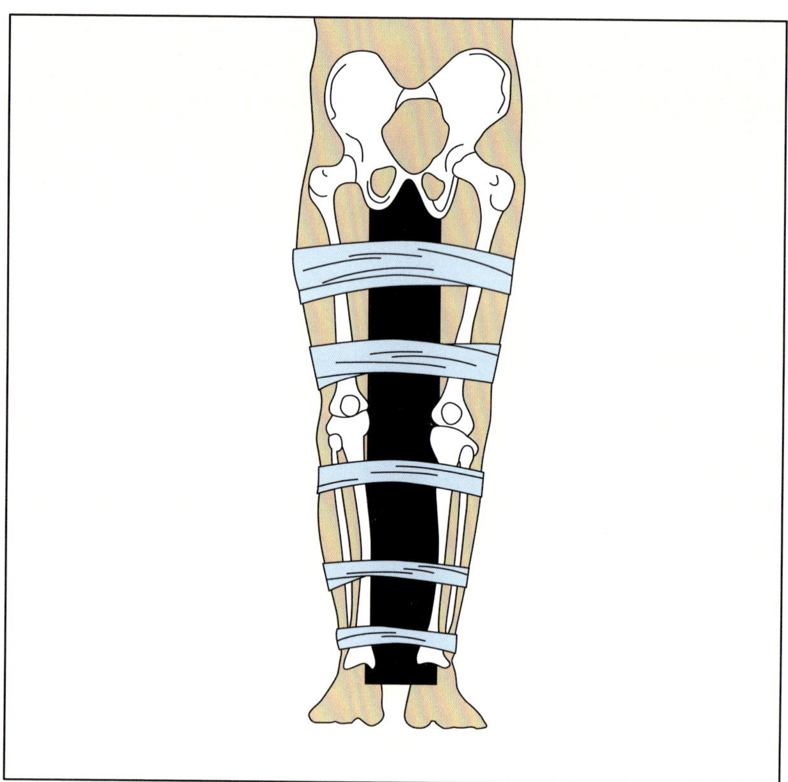

Figure 107. Anatomic leg splint

Snakebites kill very few people, less than 12 in the U.S. annually. Most deaths occur from allergic reactions or poor health.

Snakebite leaves a visible bite mark and causes pain, but try to avoid panic. A snakebite victim should lie down and remain quiet. Exertion speeds the flow of venom in the bloodstream.

Treatment

Wash the wound, immobilize the injured area, and keep it lower than the heart, if possible. Do not apply ice and do not cut the wound or apply a tourniquet. Use a snakebit kit if one is available and professional medical care could be delayed more than 30 minutes. Have the victim carried or walk very slowly when taking him or her to a medical facility.

Head and spine injuries are only a small fraction of all injuries but they cause a large portion of deaths from accidents. Motor vehicle accidents cause many head and spine injuries (fig. 108); falls are the next leading cause.

Suspect a head or spine injury if:

- The victim fell more than 5 or 6 feet (1.5 or 2 metres).
- A person is found unconscious for an unknown reason.

Snakebite

Head and Spine Injuries

Figure 108. Motor vehicle accidents cause many head and spine injuries.

- A person has been thrown from a motor vehicle.
- A person involved in a vehicle crash was not wearing a seat belt.
- A victim's hard hat was broken.
- A lightning strike was involved.

Symptoms

The victim's condition can signal a head or spine injury. Watch for a change in consciousness; a problem with breathing or vision; persistent headache, nausea, vomiting, or bruising of the head; a seizure; blood or other fluid in the ears or nose; and loss of balance.

Care

Call for an ambulance immediately if you suspect a head or spine injury. If a broken back or neck is suspected, the victim must ot be moved or allowed to move. Minimize movement of the victim's head and spine by placing the hands on both sides of the head; then gently align the victim's head and spine. Support the head in this position until medical assistance arrives. If there is resistance or if the effort causes pain, stop. Do not try to move the head if it is bent to the side. Maintain an open airway if the victim is unconscious. Control any external bleeding and keep the victim from becoming chilled or overheated.

Transporting the Injured

An injured person should not be moved until a thorough examination has been completed and all injuries are protected by dressings, splints, or other means. If the victim is severely injured, always have him or her lie down when being transported. Immobilize the victim with his or her back straight and move the entire body as a unit if the0re is a neck or spine injury.

Someone should accompany a seriously injured person being transported to a medical facility. The most accessible safe means of transportation should be used. However, it is better to use a truck bed than to try to jackknife a victim into an automobile seat.

To summarize—

General rules for emergency first aid

- At least one worker on each tour should know basic first aid and CPR.

- The rig manager should post emergency information in a designated place and provide a first-aid manual in addition to supplies such as dressings and a stretcher.

- Report all accidents to the rig manager.

- Respond to conditions that are immediately life-threatening by calling emergency services before initiating first-aid procedures.

Basic steps to take in an emergency

- Check the scene and the victim.

- Call the posted emergency number.

- Care for the victim.

Procedures for common serious emergencies

- If the victim is unconscious, check his or her breathing and pulse; check also for bleeding. Then initiate rescue breathing, if necessary.

- If the victim is bleeding severely, place a sterile dressing or clean cloth over the wound and apply pressure.

- If the victim has symptoms of shock, have him or her lie down, control external bleeding, and keep him or her warm.

- To treat a burn, remove the victim from the source of heat and cool the burn with water. Cover the burn with a sterile dressing. Do not remove anything that adheres to burned skin (such as clothing).

- To treat chemical burns to eyes and skin, wash away the chemical with water for several minutes.

- To treat electrical burns, check the victim's breathing and pulse and keep him or her warm. Do not assume if the burns on the outside seem minor that there is little damage to underlying tissues.

- If the victim has a heat-related illness, remove him or her from the heat, loosen clothing, and attempt to cool him or her with wet cloths and fans. Give small amounts of water.

- If the victim has a cold-related illness such as frostbite or hypothermia, remove him or her from the cold and attempt gradual warming.

- If the victim has symptoms of a heart attack, encourage him or her to rest and initiate CPR if he or she stops breathing.

- If the victim has a suspected fracture, immobilize the injured part with a splint unless it causes pain.

- If the victim has a suspected head or spine injury, minimize movement along the spin. Maintain an open airway if the victim is unconscious, and do not move the victim until he or she has been examined.

Glossary

accumulator *n*: 1. a vessel or tank that receives and temporarily stores a liquid used in a continuous process in a gas plant. See *drip accumulator*. 2. on a drilling rig, the storage device for nitrogen-pressurized hydraulic fluid, which is used in operating the blowout preventers. See *blowout preventer control unit*.

air gun *n*: 1. a hand tool that is powered pneumatically. 2. a chamber filled with compressed air, often used offshore in seismic exploration. As the gun is trailed behind a boat, air is released, making a low-frequency popping noise that penetrates the subsurface rock layers and is reflected by the layers. Sensitive hydrophones receive the reflections and transmit them to recording equipment on the boat.

air hoist *n*: a hoist operated by compressed air; a pneumatic hoist. Air hoists are often mounted on the rig floor and are used to lift joints of pipe and other heavy objects.

annular blowout preventer *n*: a large valve usually installed above the ram preventers, which forms a seal in the annular space between the pipe and the wellbore or, if no pipe is present, in the wellbore itself. Compare *ram blowout preventer.*

annular space *n*: the space between two concentric circles. In the petroleum industry, it is usually the space surrounding a pipe in the wellbore, the space between tubing and casing, or the space between tubing and the wellbore; sometimes termed the annulus.

annulus *n*: see *annular space.*

bail *n*: 1. a cylindrical steel bar (similar to the handle or bail of a bucket, only much larger) that supports the swivel and connects it to the hook. Sometimes, the two cylindrical bars that support the elevators and attach them to the hook are called bails or links. 2. the U-shaped handle on a nozzle used to close a valve that shuts off water flow when pushed forward. *v*: to recover bottomhole fluids, samples, or drill cuttings by lowering a cylindrical vessel called a bailer to the bottom of a well, filling it, and retrieving it.

bailer *n*: a long, cylindrical container fitted with a valve at its lower end, used to remove water, sand, mud, drilling cuttings, or oil from a well in cable-tool drilling.

beam pumping unit *n*: a machine designed specifically for sucker rod pumping. An engine or motor (prime mover) is mounted on the unit to power a rotating crank. The crank moves a horizontal member (walking beam) up and down to produce reciprocating motion. This reciprocating motion operates the pump. Compare *pump jack*.

bearing *n*: 1. an object, surface, or point that supports. 2. a machine part in which another part (such as a journal or pin) turns or slides.

bearing cap *n*: a device that is fitted around a bearing to hold or immobilize it.

bell nipple *n*: a short length of pipe (a nipple) installed on top of the blowout preventer. The top end of the nipple is flared, or belled, to guide drill tools into the hole and usually has side connections for the fill line and mud return line.

bit *n*: the cutting or boring element used in drilling oil and gas wells. The bit consists of a cutting element and a circulating element. The cutting element is steel teeth, tungsten carbide buttons, industrial diamonds, or polycrystalline diamonds (PD.C.s). These teeth, buttons, or diamonds penetrate and gouge or scrape the formation to remove it. The circulating element permits the passage of drilling fluid and utilizes the hydraulic force of the fluid stream to improve drilling rates. In rotary drilling, several drill collars are joined to the bottom end of the drill pipe column, and the bit is attached to the end of the drill collars. Drill collars provide weight on the bit to keep it in firm contact with the bottom of the hole. Most bits used in rotary drilling are roller cone bits, but diamond bits are also used extensively.

bitumastic material *n*: a compound of asphalt and filler that is used to coat metals exposed to corrosion or weathering.

blank liner *n*: liner with no perforations.

block *n*: any assembly of pulleys on a common framework; in mechanics, one or more pulleys, or sheaves, mounted to rotate on a common axis. The crown block is an assembly of sheaves mounted on beams at the top of the derrick or mast. The drilling line is reeved over the sheaves of the crown block alternately with the sheaves of the traveling block, which is hoisted and lowered in the derrick or mast by the drilling line. When elevators are attached to a hook on a conventional traveling block and drill pipe is latched in the elevators, the pipe can be raised or lowered in the derrick or mast. See *crown block, traveling block*.

blowout *n*: an uncontrolled flow of gas, oil, or other well fluids into the atmosphere. A blowout, or gusher, occurs when formation pressure exceeds the pressure applied to it by the column of drilling fluid. A kick warns of an impending blowout. See *kick*.

blowout preventer (BOP) *n*: one of several valves installed at the wellhead to prevent the escape of pressure either in the annular space between the casing and the drill pipe or in open hole (hole with no drill pipe) during drilling or completion operations. Blowout preventers on land rigs are located beneath the rig at the land's surface; on jackup or platform rigs, at the water's surface; and on floating offshore rigs, on the seafloor. See *annular blowout preventer, ram blowout preventer.*

blowout preventer control unit *n*: a device that stores hydraulic fluid under pressure in special containers and provides a method to open and close the blowout preventers quickly and reliably. Usually, compressed air and hydraulic pressure provide the opening and closing force in the unit. See *blowout preventer*. Also called an accumulator.

boom *n*: 1. a movable arm of tubular or bar steel used on some types of cranes or derricks to support the hoisting lines that carry the load. 2. a floating device used to contain oil. 3. a period of high activity in the oil industry. 4. a member hinged to the revolving superstructure and used as part of an attachment.

bottomhole pressure test *n*: a test that measures the reservoir pressure of the well, obtained at a specific depth or at the midpoint of the producing zone. A flowing bottomhole pressure test measures pressure while the well continues to flow; a shut-in bottomhole pressure test measures pressure after the well has been shut in for a specified period of time.

box *n*: the female section of a connection. See *tool joint*.

breakout cathead *n*: a device attached to the catshaft of the drawworks that is used as a power source for unscrewing drill pipe; usually located opposite the driller's side of the drawworks. See *cathead*.

breakout tongs *n pl*: tongs that are used to start unscrewing one section of pipe from another section, especially drill pipe coming out of the hole. Compare *makeup tongs*. See also *tongs*.

buddy system *n*: a method of pairing two persons for their mutual aid or protection. The buddy system is used to ensure that each crewmember is accounted for, particularly in situations where hydrogen sulfide may be encountered.

bullet perforator *n*: a tubular device that, when lowered to a selected depth within a well, fires bullets through the casing to provide holes through which the formation fluids may enter the wellbore.

bunker suit *n*: protective firefighting clothing stored on a rig that consists of a helmet, hood, coat, pants, gloves, boots, and oxygen supply.

CAA *abbr*: Clean Air Act.

C

carbon tetrachloride *n*: a liquid used for degreasing metals; it is toxic when ingested, breathed, or exposed to the skin and is a known carcinogen.

casing *n*: 1. steel pipe placed in an oil or gas well to prevent the wall of the hole from caving in, to prevent movement of fluids from one formation to another, and to improve the efficiency of extracting petroleum if the well is productive. A joint of casing may be 16 to 48 feet (4.8 to 14.6 metres) long and from 4.5 to 20 inches (11.4 to 50.8 centimetres) in diameter. Casing is made of many types of steel alloy, which vary in strength and corrosion resistance. 2. large pipe in which a carrier pipeline is contained. Casing is used when a pipeline passes under railroad rights-of-way and some roads to shield the pipeline from the unusually high load stresses of a particular location. State and local regulations identify specific locations where casing is mandatory.

casing hanger *n*: a circular device with a frictional gripping arrangement of slips and packing rings used to suspend casing from a casinghead in a well.

cathead *n*: a spool-shaped attachment on the end of the catshaft, around which rope for hoisting and moving heavy equipment on or near the rig floor is wound. See *breakout cathead, makeup cathead.*

catline *n*: a hoisting or pulling line powered by the cathead and used to lift heavy equipment on the rig. See *cathead.*

cementing *n*: the application of a liquid slurry of cement and water to various points inside or outside the casing. See *primary cementing, secondary cementing, squeeze cementing.*

CERCLA *abbr*: Comprehensive Environmental Response, Compensation, and Liability Act.

chain tongs *n pl*: a hand tool consisting of a handle and chain that resembles the chain on a bicycle. In general, chain tongs are used for turning pipe or fittings of a diameter that is larger than one a pipe wrench would fit. The chain is looped and tightened around the pipe or fitting, and the handle is used to turn the tool so that the pipe or fitting can be tightened or loosened.

cheater *n*: a length of pipe fitted over a wrench handle to increase the leverage of the wrench. Use of a larger wrench is usually preferred. Also called a snipe.

check valve *n*: a valve that permits flow in one direction only. If the gas or liquid starts to reverse, the valve automatically closes, preventing reverse movement. Commonly referred to as a one-way valve.

circulate *v*: to pass from one point throughout a system and back to the starting point. For example, drilling fluid is circulated out of the suction pit, down the drill pipe and drill collars, out the bit, up the annulus, and back to the pits while drilling proceeds.

Clean Air Act (CAA) *n*: a U.S. federal law designed to control air pollution. The EPA is the federal agency that sets and enforces emissions standards under the act.

Clean Water Act (CWA) *n*: the Federal Water Pollution Control Act of 1972, the federal law that aims to protect the integrity of U.S. waters by controlling the discharge of pollutants. The EPA is the federal agency that sets standards for the discharge of pollutants for industry.

clutch *n*: a coupling used to connect and disconnect a driving and a driven part of a mechanism, especially a coupling that permits the former part to engage the latter gradually and without shock. In the oilfield, a clutch permits gradual engaging and disengaging of the equipment driven by a prime mover. *v*: to engage or disengage a clutch.

collar *n*: 1. a coupling device used to join two lengths of pipe, such as casing or tubing. A combination collar has left-hand threads in one end and right-hand threads in the other. 2. a drill collar. See *drill collar.*

collar clamp *n*: A device that wraps around the drill collar and when closed and tightened prevents the drill collar from slipping through the rotary.

collar pipe *n*: heavy pipe used between the drill pipe and the bit in the drill stem. See *drill collar.*

come-along *n*: a manually operated device that is used to tighten guy wires or move heavy loads. Usually, a come-along is a gripping tool with two jaws attached to a ring so that, when the ring is pulled, the jaws close.

compound *n*: 1. a mechanism used to transmit power from the engines to the pump, the drawworks, and other machinery on a drilling rig. It is composed of clutches, chains and sprockets, belts and pulleys, and a number of shafts, both driven and driving. 2. a substance formed by the chemical union of two or more elements in definite proportions; the smallest particle of a chemical compound is a molecule. *v*: to connect two or more power-producing devices, such as engines, to run driven equipment, such as the drawworks.

Comprehensive Environmental Response, Compensation, and Liability Act (CERCLA) *n*: the U.S. federal law that levied a tax on the chemical and petroleum industries to provide funds for cleaning up hazardous waste sites. It also provided authority to the EPA and to states to respond directly to releases and threatened releases of hazardous substances into the environment. Also called Superfund.

confined space *n*: a space large enough and so configured that an employee can bodily enter and perform work but 1) has limited entry or exit, 2) has insufficient natural ventilation, 3) could contain hazardous materials, and 4) is not intended for continuous occupancy. Confined space may be either nonpermit or permit-required space.

confined space permit *n*: a permit used to ensure that work in a confined space is done safely. The permit sets out conditions to be met before work begins.

core *n*: 1. a cylindrical sample taken from a formation for geological analysis. Usually a conventional core barrel is substituted for the bit and procures a sample as it penetrates the formation. See *sidewall coring.* 2. the metallic, partly solid and partly molten interior of the earth, which is about 4,400 miles (7,084 kilometres) in diameter. 3. the central, axial member of a wire rope around which the strands are laid. *v*: to obtain a solid, cylindrical formation sample for analysis.

Corod® *n*: a trade name for a special form of sucker rod. Corod, or continuous rod, normally has no joints between the downhole pump and the surface. It is installed into the well by unwinding it from a reel. See *sucker rod.*

crew *n*: 1. the workers on a drilling or workover rig, including the driller, the derrickhand, and the rotary helpers. 2. any group of oilfield workers.

crown block *n*: an assembly of sheaves mounted on beams at the top of the derrick or mast and over which the drilling line is reeved. See *block, reeve the line, sheave.*

crown platform *n*: the working platform at the top of the derrick or mast that permits access to the sheaves of the crown block and provides a safe working area for service to the gin pole. See *crown block.*

CWA *abbr*: Clean Water Act.

cylinder liner *n*: a removable, replaceable sleeve that fits into a cylinder. When the sliding of the piston and rings wears out the liner, it can be replaced without the block's having to be replaced.

D

deadline *n*: the drilling line from the crown block sheave to the anchor, so called because it does not move. Compare *fastline*.

deadline anchor *n*: see *deadline tie-down anchor*.

deadline sheave *n*: the sheave on the crown block over which the deadline is reeved.

deadline tie-down anchor *n*: a device to which the deadline is attached, securely fastened to the mast or derrick substructure. Also called a deadline anchor.

deadman *n*: 1. a buried anchor to which guy wires are tied to steady the derrick, mast, stacks, and so on. 2. an anchoring point against which the winch on a boring machine for pipelining can pull.

defensive hoses *n pl*: large-size hoses used to throw large amounts of water from a distance.

degasser *n*: the device used to remove unwanted gas from a liquid, especially from drilling fluid.

deluge valve *n*: a valve that opens quickly and allows a full volume of water to flow immediately.

demister *n*: in an evaporator, a separator that removes droplets of liquid from water vapor.

derrick *n*: a large load-bearing structure, usually of bolted construction. In drilling, the standard derrick has four legs standing at the corners of the substructure and reaching to the crown block. The substructure is an assembly of heavy beams used to elevate the derrick and provide space to install blowout preventers, casingheads, and so forth. Because the standard derrick must be assembled piece by piece, it has largely been replaced by the mast, which can be lowered and raised without disassembly. Compare *mast*.

derrick floor *n*: also called the rig floor or the drill floor. See *rig floor*.

derrickhand *n*: the crewmember who handles the upper end of the drill string as it is being hoisted out of or lowered into the hole. On a drilling rig, he or she is also responsible for the circulating machinery and the conditioning of the drilling or workover fluid.

derrickman *n*: see *derrickhand*.

desander *n*: a centrifugal device for removing sand from drilling fluid to prevent abrasion of the pumps. It may be operated mechanically or by a fast-moving stream of fluid inside a special cone-shaped vessel, in which case it is sometimes called a hydrocyclone. Compare *desilter*.

desilter *n*: a centrifugal device for removing very fine particles, or silt, from drilling fluid to keep the amount of solids in the fluid at the lowest possible point. Usually, the lower the solids content of mud, the faster is the rate of penetration. The desilter works on the same principle as a desander. Compare *desander*.

die *n*: a tool used to shape, form, or finish other tools or pieces of metal. For example, a threading die is used to cut threads on pipe.

diesel fuel *n*: a light hydrocarbon mixture for diesel engines, similar to furnace fuel oil; it has a boiling range just above that of kerosene.

doghouse *n*: 1. a small enclosure on the rig floor used as an office for the driller and as a storehouse for small objects. 2. any small building used as an office, a change house, or a place for storage.

double *n*: a length of drill pipe, casing, or tubing consisting of two joints screwed together. Compare *fourble, single, thribble.*

double board *n*: the name used for the working platform of the derrickhand (the monkeyboard) when it is located at a height in the derrick or mast equal to two lengths of pipe joined together. Compare *fourble board, thribble board.*

downhole *adj, adv*: pertaining to the wellbore.

drawworks *n*: the hoisting mechanism on a drilling rig. It is essentially a large winch that spools off or takes in the drilling line and thus raises or lowers the drill stem and bit.

drill collar *n*: a heavy, thick-walled tube usually made of steel and placed between the drill pipe and the bit in the drill stem. Several drill collars are used to provide weight on the bit and to provide a pendulum effect to the drill stem. When manufactured to American Petroleum Institute (API) specifications, a drill collar joint is 30 or 31 feet (9.14 to 9.45 metres) long. The outside diameter of drill collars made to API specifications ranges from 3⅛ inches to 11 inches (79 millimetres to 275 millimetres).

driller *n*: the employee directly in charge of a drilling or workover rig and crew. The driller's main duty is operating the drilling and hoisting equipment, but this person is also responsible for the downhole condition of the well, operation of downhole tools, and pipe measurements.

driller's console *n*: a metal cabinet on the rig floor containing the controls that the driller manipulates to operate various components of the drilling rig.

driller's control panel *n*: see *driller's console.*

drilling block *n*: a lease or a number of leases of adjoining tracts of land that constitute a unit of acreage sufficient to justify the expense of drilling a wildcat.

drilling contractor *n*: an individual or group of individuals who own a drilling rig or mast and contract their services for drilling wells to a certain depth.

drilling crew *n*: a driller, a derrickhand, and two or more helpers who operate a drilling or workover rig for one tour each day.

drilling fluid *n*: circulating fluid, one function of which is to lift cuttings out of the wellbore and to the surface. It also serves to cool the bit and to counteract downhole formation pressure. Although a mixture of barite, clay, water, and other chemical additives is the most common drilling fluid, wells can also be drilled by using air, gas, water, or oil-base mud as the drilling mud. Also called circulating fluid, drilling mud. See *mud.*

drilling mud *n*: a specially compounded liquid circulated through the wellbore during rotary drilling operations. Also called drilling fluid, mud.

drilling superintendent *n*: an employee, usually of a drilling contractor, who is in charge of all drilling operations that the contractor is engaged in. Also called a toolpusher, rig manager, rig supervisor, or drilling foreman.

drill pipe slips *n pl*: see *slips.*

drillship *n*: a self-propelled floating offshore drilling unit that is a ship constructed to permit a well to be drilled from it. Although not as stable as semisubmeribles, drillships are capable of drilling exploratory wells in deep, remote waters.

drill site *n*: the location of a drilling rig.

drill stem *n*: all members in the assembly used for rotary drilling from the swivel to the bit, including the kelly, the drill pipe and tool joints, the drill collars, the stabilizers, and various specialty items. Compare *drill string*.

drill stem safety valve *n*: a special valve normally installed below the kelly. Usually, the valve is open so that drilling fluid can flow out of the kelly and down the drill stem. It can, however, be manually closed with a special wrench when necessary. In one case, the valve is closed and broken out still attached to the kelly to prevent drilling mud in the kelly from draining onto the rig floor. In another case, when kick pressure inside the drill stem exists, the drill stem safety valve is closed to prevent the pressure from escaping up the drill stem. Also called lower kelly cock, mud saver valve.

drill stem test (DST) *n*: the conventional method of formation testing. The basic drill stem test tool consists of a packer or packers, valves or ports that may be opened and closed from the surface, and two or more pressure-recording devices. The tool is lowered on the drill string to the zone to be tested. The packer or packers are set to isolate the zone from the drilling fluid column. The valves or ports are then opened to allow for formation flow while the recorders chart static pressures. A sampling chamber traps clean formation fluids at the end of the test. Analysis of the pressure charts is an important part of formation testing.

drill stem test gases *n pl*: formation gases that are produced during a drill stem test, which is the conventional method of testing formations. Although these gases typically are produced only in very small quantities and consist mainly of natural gas, they become an environmental concern when the gases fall into the extremely hazardous substance (EHS) category under SARA and are produced in sufficient quantities so that they exceed the threshold planning quantity (TPQ) limit set for that substance. Typically, wells are not set up for production operations when the drill stem test is run; consequently, gas-containment equipment is not installed or operational during the running of the test, and such gases must be released or flared.

drill string *n*: the column, or string, of drill pipe with attached tool joints that transmits fluid and rotational power from the kelly to the drill collars and the bit. Often, especially in the oil patch, the term is loosely applied to include both drill pipe and drill collars. Compare *drill stem*.

drip accumulator *n*: the device used to collect liquid hydrocarbons that condense out of a wet gas traveling through a pipeline.

drive chain *n*: a chain that propels the movement of a machine.

drum *n*: 1. a cylinder around which wire rope is wound in the drawworks. The drawworks drum is that part of the hoist on which the drilling line is wound. 2. a steel container of general cylindrical form. Some refined products are shipped in steel drums with capacities of about 50 to 55 U.S. gallons, or about 200 litres.

DST *abbr*: drill stem test.

DST tool *n*: drill stem test tool; used for formation evaluation.

dynamic positioning *n*: a method by which a floating offshore drilling rig is maintained in position over an offshore well location without the use of mooring anchors. Generally, several propulsion units, called thrusters, are located on the hulls of the structure and are actuated by a sensing system. A computer to which the system feeds signals directs the thrusters to maintain the rig on location.

elevator bails *n pl*: see *elevator links*.

elevator links *n pl*: cylindrical bars that support the elevators and attach them to the hook. Also called elevator bails.

elevators *n pl*: on conventional rotary rigs and top-drive rigs, hinged steel devices with manual operating handles that crewmembers latch onto a tool joint (or a sub). Since the elevators are directly connected to the traveling block or to the integrated traveling block in the top drive, when the driller raises or lowers the block or the top-drive unit, the drill pipe is also raised or lowered.

Environmental Protection Agency (EPA) *n*: the U.S. federal agency that sets and enforces regulations to protect the environment and human health. The broad authority of the EPA is granted by many pieces of federal legislation, including the Clean Air Act, the Clean Water Act, the Endangered Species Protection Act, and the Comprehensive Environmental Response, Compensation, and Liability Act (CERCLA), or Superfund. Address: 1200 Pennsylvania Avenue, NW; Washington, D.C. 20460; 202-272-0167; www.epa.gov.

EPA *abbr*: Environmental Protection Agency.

ergonomic *adj*: intended to maintain comfort while avoiding stress and injury. Ergonomic designs of equipment take into account the human form so that human operatives can use the equipment efficiently without strain. Ergonomic body positions and ways of movement also reduce strain or injury in the workplace.

fastline *n*: the end of the drilling line that is affixed to the drum or reel of the drawworks, so called because it travels with greater velocity than any other portion of the line. Compare *deadline*.

field office manager *n*: the individual responsible for the contractor's financial affairs on a pipeline spread. The field office manager oversees billing arrangements, payroll, and other money-related matters.

field superintendent *n*: an employee of an oil company who is in charge of a particular oil or gas field from which the company is producing.

field support personnel *n*: in pipeline construction, the mechanics, parts and warehouse workers, truck drivers, and others who service the machinery that actually lays pipe.

fingerboard *n*: a rack that supports the tops of the stands of pipe being stacked in the derrick or mast. It has several steel fingerlike projections that

173

form a series of slots into which the derrickhand can place a stand of drill pipe after it is pulled out of the hole and removed from the drill string.

fish *n*: an object that is left in the wellbore during drilling or workover operations and that must be recovered before work can proceed. It can be anything from a piece of scrap metal to a part of the drill stem. *v*: 1. to recover from a well any equipment left there during drilling operations, such as a lost bit, drill collar, or part of the drill string. 2. to remove from an older well certain pieces of equipment (such as packers, liners, or screen liner) to allow reconditioning of the well.

fishing tool *n*: a tool designed to recover equipment lost in a well.

fishing-tool operator *n*: the person (usually a service company employee) in charge of directing fishing operations.

flag *n*: 1. a piece of cloth, rope, or nylon strand used to mark the wireline when swabbing or bailing. 2. an indicator of wind direction used during drilling or workover operations when hydrogen sulfide (sour) gas may be encountered. *v*: 1. to signal or attract attention. 2. in swabbing or bailing, to attach a piece of cloth to the wireline to enable the operator to estimate the position of the swab or bailer in the well.

flange *n*: a projecting rim or edge (as on pipe fittings and openings in pumps and vessels) usually drilled with holes to allow bolting to other flanged fittings.

floor crew *n*: those workers on a drilling or workover rig who work primarily on the rig floor. See *rotary helper*.

floorhand *n*: see *rotary helper*.

floorman *n*: see *rotary helper*.

fog *n*: a wide, fine spray from a nozzle.

formation pressure *n*: the force exerted by fluids in a formation and recorded in the hole at the level of the formation with the well shut in. Also called reservoir pressure or shut-in bottomhole pressure.

formation testing *n*: the gathering of pressure data and fluid samples from a formation to determine its production potential before choosing a completion method. Formation testing tools include formation testers and drill stem test tools.

fourble *n*: a section of drill pipe, casing, or tubing consisting of four joints screwed together. Compare *double, single, thribble*.

fourble board *n*: the name used for the working platform of the derrickhand, or the monkeyboard, when it is located at a height in the derrick equal to approximately four lengths of pipe joined together. Compare *double board, thribble board*.

G **gallonage** *n*: the amount of liquid a fire-fighting device delivers in U.S. gallons.

gauge *n*: a device (such as a pressure gauge) used to measure some physical property.

GHS *abbr*: Globally Harmonized System of Classification and Labelling of Chemicals.

gin pole *n*: a pole (usually single) with guy wires and used with block and tackle to hoist equipment. On a drilling rig, the gin pole is typically secured to the mast or derrick above the monkeyboard.

gin pole truck *n*: a truck equipped with hoisting equipment and a pole or arrangement of poles for use in lifting heavy machinery.

girt *n*: one of the horizontal braces between the legs of a derrick.

GL *abbr*: ground level; used in drilling reports.

gland *n*: a device used to form a seal around a reciprocating or rotating rod (as in a pump) to prevent fluid leakage. Specifically, the movable part of a stuffing box by which the packing is compressed.

gland packing *n*: material placed around a gland to effect a seal around a reciprocating or rotating rod.

gland-packing nut *n*: a threaded device whose sides are arranged so that a wrench can be fitted onto them and used to retain the gland packing in place around a rod. See *gland packing*.

Globally Harmonized System of Classification and Labelling of Chemicals (GHS) *n*: a system designed by the United Nations for harmonizing the international classification and labeling of chemicals. OSHA's Hazard Communication Standard was amended in 2012 to align requirements in the U.S. with this international system. The new labels and safety data sheets (SDSes) will replace the labels and material safety data sheets (MSDSes) formerly mandated by the HCS. Chemical manufacturers, distributors, and importers will be required to comply with the new standard on June 1, 2015, but may comply earlier. See *material safety data sheet, safety data sheet, Hazard Communication Standard*.

gooseneck *n*: the curved connection between the rotary hose and the swivel. See *swivel*.

gun-perforate *v*: to create holes in casing and cement that run through a productive formation. A common method of completing a well is to set casing through the oil-bearing formation and cement it. A perforating gun is then lowered into the hole and fired to detonate high-powered jets or shoot steel projectiles (bullets) through the casing and cement and into the pay zone. The formation fluids flow out of the reservoir through the perforations and into the wellbore. See *jet-perforate, perforating gun*.

gusher *n*: an oilwell that has come in with such great pressure that the oil jets out of it like a geyser. In reality, a gusher is a blowout and is extremely wasteful of reservoir fluids and drive energy. In the early days of the oil industry, gushers were common and many times were the only indication that a large reservoir of oil and gas had been struck. See *blowout*.

guying system *n*: the system of guy lines and anchors used to brace a rig.

guy line *n*: a wireline attached to a mast, derrick, or offshore platform to stabilize it. See *load guy line, wind guy line*.

guy line anchor *n*: a buried weight or anchor to which a guy line is attached. See *deadman*.

guy rope *n*: a supporting rope that maintains a constant distance between the points of attachment to the two components connected by the rope.

guy wire *n*: a rope or cable used to steady a mast or pole.

H

hard hat *n*: a hard plastic helmet worn by oilfield workers to minimize the danger of being injured by falling objects.

Hazard Communication Standard (HAZCOM, HCS, or the "Employee Right to Know") *n*: an OSHA standard that guarantees employees the right to know about chemical hazards on the job and how to protect themselves from those hazards. Under HCS, all manufacturers and employers must prepare a written hazard communication program, prepare a list of all hazardous materials in the workplace, label all containers of hazardous materials in the workplace, collect and maintain a safety data sheet (SDS) for each hazardous material present in the workplace, and provide employee training on specific topics related to hazardous substances.

hazardous *adj*: involving or exposing one to risk. The lists of materials or types of waste that are considered hazardous vary from agency to agency and from regulation to regulation. Hazardous materials in transport are regulated by DOT, hazardous substances in the workplace are regulated by OSHA, hazardous waste is regulated under the EPA's RCRA, toxic substances are regulated under the EPA's TSCA, and hazardous air pollutants are regulated under the EPA's CAA. The hazardous list for that regulation is tailored to that purpose.

hazardous airborne pollutant (HAP) *n*: under the CAA, air emissions that are immediately hazardous to human health or that cause cancer, gene mutation, or reproductive harm. Due to its harmful nature, the allowable emission of a hazardous airborne pollutant is much lower than that for a conventional, or criteria, pollutant.

hazardous chemical *n*: 1. under the HCS, any chemical that is a physical hazard or a health hazard. 2. under SARA, any hazardous chemical as defined under 29 CFR 1910.1200, except those that are regulated by other agencies or laws.

hazardous materials (HAZMAT) *n pl*: as defined by the U.S. Department of Transportation (DOT), substances or materials in quantities or forms that may pose an unreasonable risk to health, safety, or property when stored, transported, or used in commerce.

hazardous materials specialist level *n*: a training level achieved by any employee who has been HAZWOPER trained to assist and support a hazardous materials technician in making certain emergency action decisions. The duties of hazardous materials specialists parallel those of the technician, but require a more directed or specific knowledge of the various substances they may be called on to contain. The hazardous materials specialist also acts as the site liaison with federal, state, local, and other government authorities in regard to site activities.

hazardous materials technician level *n*: a training level achieved by any employee who has been HAZWOPER trained to take an aggressive role in emergency response. Technicians are trained to take certain actions that deal directly with stopping a release, such as approaching the point of release to plug, patch, or stop the release otherwise.

hazardous substance *n*: 1. any substance designated under the CWA or CERCLA as posing a threat to waterways and the environment when released. 2. under CERCLA, any substance designated in 40 CFR 302.

hazardous waste *n*: 1. any solid waste—solid, liquid, semisolid, or contained gaseous material—resulting from industrial, commercial, mining, or agricultural operations or from community activities and that has certain characteristics of a hazard (being ignitable, corrosive, reactive, or toxic), that is listed as a waste from specific or nonspecific sources, or that is a listed commercial chemical product or manufacturing intermediate that is sometimes discarded. 2. under the RCRA, discarded materials regulated by the EPA because of public health and safety concerns. 3. under the HAZWOPER, a waste or combination of wastes as defined in 40 CFR 261.3, or those substances defined as hazardous wastes in 49 CFR 171.8. Under CERCLA, those wastes listed in 40 CFR 261.3.

hazardous waste generator *n*: an operator that produces hazardous or acute hazardous waste. If the quantity of waste exceeds EPA minimums under the RCRA, the operator must obtain a generator identification number and must meet other RCRA requirements. The hazardous waste generator must place hazardous wastes in proper containers, label the containers, ensure safe handling of the material, manifest shipments to licensed disposal sites, and report discrepancies in waste shipments to the EPA.

hazardous waste manifest *n*: a document that identifies the waste and all parties responsible for it while it is being shipped.

Hazardous Waste Operations and Emergency Response Standard (HAZWOPER) *n*: an OSHA standard that is concerned primarily with worker safety in emergency response situations. HAZWOPER requires employers to protect the safety and health of three specific groups of workers: those involved in emergency response or cleanup at hazardous waste sites; those involved in emergency response at treatment, storage, and disposal (TSD) sites; and those involved in emergency response to incidents involving hazardous substances.

hazard warning *n*: as required by OSHA, any words, pictures, symbols, or combination thereof appearing on a label or other appropriate form of warning that convey the hazard(s) of the chemical(s) in the container(s).

HAZCOM *abbr*: Hazard Communication Standard.

HAZMAT *abbr*: hazardous material.

HAZMAT team *n*: the designated and trained personnel who respond to hazardous material incidents.

HAZWOPER *abbr*: Hazardous Waste Operations and Emergency Response Standard.

HCN *form*: hydrogen cyanide.

HCS *abbr*: Hazard Communication Standard.

head well puller *n*: crew chief.

health hazard *n*: as defined by OSHA, the potential human health hazard associated with contact with materials or substances. A chemical that is listed as a health hazard is a chemical for which there is statistically significant evidence based on at least one study conducted in accordance with established scientific principles and that shows acute or chronic health effects may occur in exposed employees.

health, safety, and environment (HSE) *n*: a phrase describing a company's concern for the health and safety of its employees, as well as the environment in which they work.

helmet *n*: a protective enclosure for a diver's entire head. It is part of the life-support system and also contains a communications system.

hoist *n*: 1. an arrangement of pulleys and wire rope or chain used for lifting heavy objects; a winch or similar device. 2. the drawworks. *v*: to raise or lift.

hoisting *n*: the process of lifting.

hook *n*: a large, hook-shaped device from which the swivel is suspended. It is designed to carry maximum loads ranging from 100 to 650 tons (90 to 590 tonnes) and turns on bearings in its supporting housing. A strong spring within the assembly cushions the weight of a stand (90 feet, or about 27 metres) of drill pipe, thus permitting the pipe to be made up and broken out with less damage to the tool joint threads. Smaller hooks without the spring are used for handling tubing and sucker rods. See also *stand, swivel*.

hot work permit *n*: a permit used to ensure that welding and cutting on the rig site is done safely. Also called a safe work permit.

housing *n*: something that covers or protects—the casing for a moving mechanical part, for example.

HSE *abbr*: health, safety, and environment.

hull *n*: the framework of a vessel including all decks, plating, and columns, but excluding machinery.

I

IADC *abbr*: International Association of Drilling Contractors.

ignition temperature *n*: the temperature at which a particular vapor burns.

in-line proportioner *n*: a Venturi tube that draws foam concentrate into a water stream at a predetermined proportion.

International Association of Drilling Contractors (IAD.C.) *n*: an organization of drilling contractors that sponsors or conducts research on education, accident prevention, drilling technology, and other matters of interest to drilling contractors and their employees. Its official publication is *Drilling Contractor*. Address: Box 4287; Houston, TX 77210; 713-292-1945; fax 713-292-1946; www.iadc.org.

J

jack *n*: 1. an oilwell pumping unit that is powered by an internal-combustion engine, electric motor, or rod line from a central power source. The walking beam of the pumping jack provides reciprocating motion to the pump rods of the well. See *walking beam*. 2. a device that is manually operated to turn an engine over for starting. *v*: to raise or lift.

jack board *n*: a device used to support the end of a length of pipe while another length is being screwed onto the pipe. Sometimes referred to as a stabbing jack.

jackhammer *n*: 1. a rock drill that is pneumatically powered and usually held by the operator. 2. an air hammer.

jackknife mast *n*: an open-sided tower made of structural steel that is raised vertically by special lifting tackle attached to the traveling block. See *mast*. Compare *standard derrick*.

jackknife rig *n*: a drilling rig that has a jackknife mast instead of a standard derrick.

jackup *n*: a jackup drilling rig.

jackup drilling rig *n*: a mobile bottom-supported offshore drilling structure with columnar or open-truss legs that support the deck and hull. When positioned over the drilling site, the bottoms of the legs rest on the seafloor. A jackup rig is towed or propelled to a location with its legs up. Once the legs are firmly positioned on the bottom, the deck and hull height are adjusted and leveled. Also called self-elevating drilling unit.

jet-perforate *v*: to create holes through the casing with a shaped charge of high explosives instead of a gun that fires projectiles. The loaded charges are lowered into the hole to the desired depth. Once detonated, the charges emit short, penetrating jets of high-velocity gases that make holes in the casing and cement for some distance into the formation. Formation fluids then flow into the wellbore through these perforations. See *bullet perforator; gun-perforate*.

joint *n*: in drilling, a single length (from 16 feet to 45 feet, or 5 metres to 14.5 metres, depending on its range length) of drill pipe, drill collar, casing, or tubing that has threaded connections at both ends. Several joints screwed together constitute a stand of pipe.

kelly bushing (KB) *n*: a special device placed around the kelly that mates with the flat sides of the kelly and fits into the master bushing of the rotary table. The kelly bushing is designed so that the kelly is free to move up or down through it. The bottom of the bushing may be shaped to fit the opening in the master bushing or it may have pins that fit into the master bushing. In either case, when the kelly bushing is inserted into the master bushing and the master bushing is turned, the kelly bushing also turns. Since the kelly bushing fits onto the kelly, the kelly turns, and since the kelly is made up to the drill stem, the drill stem turns. Also called the drive bushing. See *master bushing*.

K

kelly bushing lock assembly *n*: a feature on a four-pin kelly bushing installed on floating offshore rigs (which employ a conventional rotary table assembly) that secures the kelly bushing's pins to the master bushing's corresponding drive holes so that the kelly bushing will not separate from (lift off of) the master bushing as the rig heaves up and down with wind and wave motion. See *kelly bushing, master bushing*.

kelly bushing rollers *n pl*: rollers in the kelly bushing that roll against the flat sides of the kelly and allow it to move freely upward or downward. Also called drive rollers.

kelly bypass *n*: a system of valves and piping that allows drilling fluid to be circulated without the use of the kelly.

kelly cock *n*: a valve installed at one or both ends of the kelly. When a high-pressure backflow occurs inside the drill stem, the valve is closed to keep pressure off the swivel and rotary hose.

kelly flat *n*: one of the flat sides of a kelly. Also called a flat.

kelly hose *n*: see *rotary hose*.

kelly spinner *n*: a pneumatically operated device mounted on top of the kelly that, when actuated, causes the kelly to turn or spin. It is useful when the kelly or a joint of pipe attached to it must be spun up—that is, rotated rapidly for being made up.

key *n*: 1. a hook-shaped wrench that fits the square shoulder of a sucker rod and is used when rods are pulled or run into a pumping oilwell. They are usually used in pairs; one key backs up and the other breaks out or makes up the rod. Also called a rod wrench. 2. a slender strip of metal that is used to fasten a wheel or a gear onto a shaft. The key fits into slots in the shaft and in the wheel or gear. *v*: to use a cotter key to prevent a nut from coming loose from a bolt or a stud.

keyseat *n*: 1. an undergauge channel or groove cut in the side of the borehole and parallel to the axis of the hole. A keyseat results from the rotation of pipe on a sharp bend in the hole. 2. a groove cut parallel to the axis in a shaft or a pulley bore.

kick *n*: an entry of water, gas, oil, or other formation fluid into the wellbore during drilling. It occurs because the pressure exerted by the column of drilling fluid is not great enough to overcome the pressure exerted by the fluids in the formation drilled. If prompt action is not taken to control the kick or kill the well, a blowout may occur.

kick fluids *n pl*: oil, gas, water, or any combination of those that enters the borehole from a permeable formation.

L

landman *n*: a person in the petroleum industry who negotiates with landowners for oil and gas leases, options, minerals, and royalties and with producers for joint operations relative to production in a field. Also called a leaseman.

land rig *n*: any drilling rig that is located on dry land. Compare *offshore rig*.

liner *n*: 1. a string of pipe used to case open hole below existing casing. A liner extends from the setting depth up into another string of casing, usually overlapping about 100 feet (30.5 metres) above the lower end of the intermediate casing or the oil string. Liners are nearly always suspended from the upper string by a hanger device. 2. a relatively short length of pipe with holes or slots that is placed opposite a producing formation. Usually, such liners are wrapped with specially shaped wire that is designed to prevent the entry of loose sand into the well as it is produced. They are also often used with a gravel pack. 3. in jet perforation guns, a conically shaped metallic piece that is part of a shaped charge. It increases the efficiency of the charge by increasing the penetrating ability of the jet. 4. a replaceable tube that fits inside the cylinder of an engine or a pump. See *cylinder liner*.

line spooler *n*: a device fitted on the drawworks and used to cause the fastline to reverse its direction on the drawworks drum or spool when a layer of line is completed on the drum and the next layer is started.

liquefied petroleum gas (LPG) *n*: a mixture of heavier, gaseous, paraffinic hydrocarbons, principally butane and propane. These gases, easily liquefied at moderate pressure, may be transported as liquids and converted to gases on release of the pressure. Thus, liquefied petroleum gas is a portable source of thermal energy that finds wide application in areas where it is impractical to distribute natural gas. It is also used as a fuel for internal-combustion engines and has many industrial and domestic uses. Principal sources of LPG are natural and refinery gas, from which the liquefied petroleum gases are separated by fractionation.

liquefied refinery gas (LRG) *n*: liquid propane or butane produced by a crude oil refinery. It may differ from LPG in that propylene and butylene may be present.

load binder *n*: A device used to tighten a chain or to hold two chains together, often used to tie a load of pipe onto a flat-bed trailer. It consists of two chains or cables, a latching device, and two hooks that fasten onto a chain or another object. Also called boomer. Rachet-type binders are replacing the older hand-thrown models.

loader *n*: the individual who handles the filling of tank cars, ships, barges, or transport trucks.

load guy *n*: see *guy line*.

load guy line *n*: the wireline attached to a mast or derrick to provide the main support for the structure. Compare *wind guy line*.

log *n*: a systematic recording of data, such as a driller's log, mud log, electrical well log, or radioactivity log. Many different logs are run in wells to discern various characteristics of downhole formation. *v*: to record data.

lost-time injury (LTI) *n*: a nonfatal injury that causes a loss of time from work—in other words, causes an employee to have to leave his or her duties for a time to recover. Lost-time injuries have a health cost for the person injured and a financial cost for the employer.

LPG *abbr*: liquefied petroleum gas.

LRG *abbr*: liquefied refinery gas.

M

make hole *v*: to deepen the hole made by the bit—to drill ahead and to run casing or pipe.

makeup *adj*: added to a system (for example, makeup water used in mixing mud).

make up a joint *v*: to screw a length of pipe into another length of pipe.

makeup cathead *n*: a device that is attached to the shaft of the drawworks and is used as a power source for screwing together joints of pipe. It is usually located on the driller's side of the drawworks. Also called spinning cathead. See *cathead*.

makeup tongs *n pl*: tongs used for screwing one length of pipe into another for making up a tool joint. Compare *breakout tongs*. See also *tongs*.

makeup water *n*: the water used as a base in water-based drilling muds. It may be fresh, brackish, or salty.

mandrel *n*: a cylindrical bar, spindle, or shaft around which other parts are arranged or attached or that fits inside a cylinder or tube.

manhole *n*: a hole in the top or side of a tank through which a person can enter.

mast *n*: a portable derrick that is capable of being raised as a unit, as distinguished from a standard derrick, which cannot be raised to a working position as a unit. For transportion by land, the mast can be divided into two or more sections to avoid excessive length extending from truck beds on the highway. Oil workers and manufacturers often use the words "mast" and "derrick" interchangeably. Compare *derrick*.

master bushing *n*: a device that fits into the rotary table to accommodate the slips and drive the kelly bushing so that the rotating motion of the rotary table can be transmitted to the kelly. Also called rotary bushing. See *kelly bushing, slips*.

material safety data sheet (MSDS) *n*: a document providing information on potential health hazards of chemicals or other substances to workers who handle them. These sheets mandated by the Hazard Communication Standard and enforced by OSHA also provide information on handling the substances and identify the manufacturer or distributor. As of June 1, 2015, OSHA will require safety data sheets (SDSes) that align with the Globally Harmonized System of Classification and Labelling of Chemicals (GHS). See *safety data sheet, Globally Harmonized System of Classification and Labelling of Chemicals, Hazard Communication Standard*.

monkeyboard *n*: the derrickhand's working platform. As pipe or tubing is run into or out of the hole, the derrickhand must handle the top end of the pipe, which may be as high as 90 feet (27 metres) or higher in the derrick or mast. The monkeyboard provides a small platform to raise the derrickhand to the proper height for handling the top of the pipe.

motorhand *n*: the crewmember on a rotary drilling rig, usually the most experienced rotary helper, who is responsible for the care and operation of drilling engines. Also called motorhand.

motorman *n*: see *motorhand*.

MSDS *abbr*: material safety data sheet.

mud *n*: the liquid circulated through the wellbore during rotary drilling and workover operations. In addition to bringing cuttings to the surface, drilling mud cools and lubricates the bit and the drill stem, protects against blowouts by holding back subsurface pressures, and deposits a mud cake on the wall of the borehole to prevent loss of fluids to the formation. Although it originally was a suspension of earth solids (especially clays) in water, the mud used in modern drilling operations is a more complex, three-phase mixture of liquids, reactive solids, and inert solids. The liquid phase may be fresh water, diesel oil, or crude oil and may contain one or more conditioners. See *drilling fluid*.

mud engineer *n*: an employee of a drilling fluid supply company whose duty it is to test and maintain the drilling mud properties that are specified by the operator.

mud hose *n*: see *rotary hose*.

mud hound *n*: see *mud engineer*.

mud logger *n*: an employee of a mud logging company who performs the logging.

mud logging *n*: recording of the information derived from the mud and cuttings circulated out of the wellbore. A sample of mud is diverted through a gas-detecting device and analyzed for gas content. Cuttings are examined under ultraviolet light for the presence of oil or gas. Mud logging is usually carried out in a portable laboratory set up at the well site.

mud man *n*: see *mud engineer*.

mud pit *n*: originally, an open pit dug in the ground to hold drilling fluid or waste materials discarded after the treatment of drilling mud. For some drilling operations, mud pits are used for suction to the mud pumps, settling of mud sediments, and storage of reserve mud. Steel tanks are much more commonly used for these purposes now, but they are still usually referred to as pits, except offshore, where "mud tanks" is preferred.

mud pump *n*: a large, high-pressure reciprocating pump used to circulate the mud on a drilling rig. A typical mud pump is a single- or double-acting, two- or three-cylinder piston pump whose pistons travel in replaceable liners and are driven by a crankshaft actuated by an engine or a motor. Also called a slush pump.

mud tank *n*: one of a series of open tanks usually made of steel plate and through which the drilling mud is cycled to remove sand and fine sediments. Additives are mixed with the mud in the tanks, and the fluid is temporarily stored there before being pumped back into the well. Modern rotary drilling rigs are generally provided with three or more tanks, fitted with built-in piping, valves, and mud agitators. Also called mud pits.

mud weight *n*: a measure of the density of a drilling fluid expressed as pounds per gallon, pounds per cubic foot, or kilograms per cubic metre. Mud weight is directly related to the amount of pressure the column of drilling mud exerts at the bottom of the hole.

NACE International *n*: an organization whose function is to establish standards and recommended practices for the field of corrosion control. Address: 1440 South Creek Drive; Houston, TX 77084; 281-228-6200; fax 281-228-6300; www.nace.org.

National Ambient Air Quality Standards (NAAQS) *n pl*: standards listed under the CAA for six major, or criteria, pollutants: ozone, carbon monoxide, sulfur dioxide, lead, nitrogen dioxide, and particulate matter. Areas that exceed the recommended levels for these pollutants are nonattainment areas, or areas with poor air quality; areas that meet or fall below the NAAQS levels are attainment areas, or areas with good air quality.

National Emissions Standards for Hazardous Airborne Pollutants (NESHAP) *n pl*: emissions standards set forth under the CAA for airborne pollutants that are immediately hazardous to human health or that cause cancer, gene mutations, or reproductive harm.

National Environmental Policy Act (NEPA) *n*: a Congressional act that forms the basic national charter for protection of the environment. It ensures that no agency of the federal government will take action that will significantly affect the quality of the human environment.

National Institute for Occupational Safety and Health (NIOSH) *n*: an organization established by the OSH Act within the Department of Health and Human Services to develop and establish recommended occupational safety and health standards and to conduct research. The OSH Act requires NIOSH to publish an annual listing of all known toxic substances and the concentrations at which toxicity is known to occur.

National Pollutant Discharge Elimination System (NPDES) *n*: a permit system set up under the CWA and implemented by qualified state governments to regulate pollutant discharges. NPDES permits specify the types of control equipment required and the discharges allowed for each facility. The permits specify levels of performance, and failure to achieve these levels must be reported. All permits can be reviewed by the EPA, the Army Corps of Engineers, and the Fish and Wildlife Service.

National Response Center (NRC) *n*: the USCG headquarters for emergency incidents. The NRC is operated 24 hours a day by the USCG in cooperation with 14 other federal agencies. Whenever a hazardous material release occurs in American waters, it must be reported to the NRC under federal law. Address: 2100 2nd Street, Room 2611; Washington, D.C. 20593; 800-424-8802; www.uscg.mil.

National Strike Force Coordination Center (NSFCC) *n*: the USCG national response system headquartered in Elizabeth City, North Carolina. The NSFCC is charged with maintaining a comprehensive list of spill removal resources, personnel, and equipment that is available worldwide and to make that list available to federal and state agencies and the public. The NSFCC provides technical assistance, equipment, and other resources required by the federal on-scene coordinator; coordinates the use of private and public personnel and equipment to remove a worst-case discharge and mitigate or prevent a substantial threat of such discharge; administers the coast guard strike teams; and provides technical assistance in preparing area contingency plans. The NSFCC maintains on file all area contingency plans and is required to review each of those plans that affects its responsibility. Address: 1461 North Road Street; Elizabeth City, NC 27909; 252-331-6000; www.uscg.mil.

National Stripper Well Association (NSWA) *n*: an association of producers of wells that yield ten barrels or less of crude petroleum a day. Address: Box 18336; Oklahoma City, OK 73154; 405-228-4112; nswa.us.

neck down *v*: to taper to a reduced diameter. A pipe becomes necked down when it is subjected to excessive longitudinal stress.

needle valve *n*: a globe valve that contains a sharp-pointed, needlelike plug that is driven into and out of a cone-shaped seat to accurately control a relatively small rate of fluid flow. In a fuel injector, the fuel pressure forces the needle valve off its seat to allow injection.

NEPA *abbr*: National Environmental Policy Act.

NESHAP *abbr*: National Emissions Standards for Hazardous Airborne Pollutants.

O

Occupational Safety and Health (OSH) Act of 1970 *n*: a Congressional act passed "to assure so far as possible every working man and woman in the nation safe and healthful working conditions and to preserve our human resources." The OSH Act establishes the Occupational Safety and Health Administration as a federal agency and authorizes it to promulgate, modify, or revoke occupational safety and health standards.

Occupational Safety and Health Administration (OSHA) *n*: a U.S. federal government agency that sets and enforces regulations concerning safe and healthy practices in the workplace. Address: U. S. Department of Labor; 200 Constitution Avenue, NW; Washington, D.C. 20210; 202-523-1452; www.osha.gov.

offshore *n*: the geographic area that lies seaward of the coastline. In general, the term "coastline" means the line of ordinary low water along the portion of the coast that is in direct contact with the open sea or the line marking the seaward limit of inland waters.

offshore drilling *n*: drilling for oil or gas in an ocean, gulf, or sea, usually on the Outer Continental Shelf. A drilling unit for offshore operations may be a mobile floating vessel with a ship or barge hull, a semisubmersible or submersible base, a self-propelled or towed structure with jacking legs (a jackup drilling rig), or a permanent structure used as a production platform when drilling is completed. In general, wildcat wells are drilled from mobile floating vessels or from jackups, while development wells are drilled from platforms or jackups.

offshore rig *n*: any of various types of drilling structures designed to drill wells in oceans, seas, bays, and gulfs. Offshore rigs include platforms, jackup drilling rigs, semisubmersible drilling rigs, submersible drilling rigs, and drillships. Compare *land rig*.

oil-base mud *n*: a drilling or workover fluid in which oil is the continuous phase and which contains 2% to 5% water. This water is spread out, or dispersed, in the oil as small droplets.

operator *n*: the person or company, either proprietor or lessee, actually operating an oilwell or lease, generally the oil company that engages the drilling, service, and workover contractors.

OSHA *abbr*: Occupational Safety and Health Administration.

OSH Act *abbr*: Occupational Safety and Health Act of 1970.

P

packed-hole assembly *n*: a bottomhole assembly consisting of stabilizers and large-diameter drill collars arranged in a particular configuration to maintain the drift angle and direction of a hole.

perforate *v*: to pierce the casing wall and cement of a wellbore to provide holes through which formation fluids may enter or to provide holes in the casing so that materials may be introduced into the annulus between the casing and the wall of the borehole. Perforating is accomplished by lowering into the well a perforating gun, or perforator, which fires electrically detonated bullets or shaped charges. See *perforating gun.*

perforated completion *n*: 1. a well completion method in which the producing zone or zones are cased through, cemented, and perforated to allow fluid to flow into the wellbore. 2. a well completed by this method.

perforated liner *n*: a liner that has had holes shot in it by a perforating gun. See *liner.*

perforated pipe *n*: sections of pipe (such as casing, liner, and tail pipe) in which holes or slots have been cut before it is set.

perforating gun *n*: a device fitted with shaped charges or bullets that is lowered to the desired depth in a well and fired to create penetrating holes in casing, cement, and formation.

pin *n*: 1. the male threaded section of a tool joint. 2. on a bit, the threaded bit shank. 3. one of the pegs that is fitted on each side into the link plates (side bars) of a chain link of roller chain and that serve as the stable members onto which bushings are press-fitted and around which rollers move. See *wrist pin.*

pinch bar *n*: a steel lever with a pointed projection at one end; it is used to lift a heavy load.

pinch point *n*: any point at which it is possible for a person or part of a person's body to be caught between moving parts of a machine, between the moving and stationary parts of a machine, or between material and any part of the machine.

pipe gang *n*: in pipeline construction, the workers responsible for positioning the pipe, aligning it, and making the initial welds. The pipe gang sets the pace that determines the progress of the rest of the spread.

pipe hanger *n*: 1. a circular device with a frictional gripping arrangement used to suspend casing and tubing in a well. 2. a device used to support a pipeline.

pipe rack *n*: a horizontal support for tubular goods.

pipe-racking fingers *n pl*: extensions within a pipe rack for keeping individual pipes separated.

platform *n*: see *platform rig.*

platform jacket *n*: a support that is firmly secured to the ocean floor and to which the legs of a platform are anchored.

platform rig *n*: an immobile offshore structure from which development wells are drilled and produced. Platform rigs may be built of steel or concrete and may be rigid or compliant. Rigid platform rigs, which rest on the seafloor, include the caisson-type platform, the concrete gravity platform, and the steel-jacket platform. Compliant platform rigs, which are used in deeper waters and yield to water and wind movements, include the guyed-tower platform, the tension-leg platform, and the compliant piled tower platform.

pneumatic *adj*: operated by air pressure.

pneumatic control *n*: a control valve that is actuated by air. Several pneumatic controls are used on drilling rigs to actuate and control rig components (such as clutches, hoists, engines, and pumps).

pneumatic line *n*: any hose or line, usually reinforced with steel, that conducts air from an air source (such as a compressor) to a component that is actuated by air (such as a clutch).

pneumatic tube fire detector *n*: a system that can detect fires in open spaces, consisting of a length of flexible plastic or metal tubing that forms a loop around the outside of the structure it protects.

pressure relief valve *n*: a valve that opens at a preset pressure to relieve excessive pressures within a vessel or line. Also called a pop valve, relief valve, safety valve, or safety relief valve.

primary cementing *n*: the cementing operation that takes place immediately after the casing has been run into the hole. It provides a protective sheath around the casing, segregates the producing formation, and prevents the undesirable migration of fluids.

prime mover *n*: an internal-combustion engine or a turbine that is the source of power for driving a machine or machines.

production rig *n*: a portable servicing or workover outfit, usually mounted on wheels and self-propelled. A well-servicing unit consists of a hoist and engine mounted on a wheeled chassis with a self-erecting mast. A workover rig is basically the same, but also has a substructure with rotary, pump, pits, and auxiliaries to permit handling and working a drill string.

pump *n*: a device that increases the pressure on a fluid or raises it to a higher level. Various types of pumps include the bottomhole pump, centrifugal pump, hydraulic pump, jet pump, mud pump, reciprocating pump, rotary pump, sucker rod pump, and submersible pump.

pump house *n*: a building that houses the pumps, engines, and control panels at a pipeline gathering station or trunk station.

pump jack *n*: a surface unit similar to a pumping unit but having no individual power plant. Usually, several pump jacks are operated by pull rods or cables from one central power source. Commonly but erroneously, beam pumping units are called pump jacks. Compare *beam pumping unit*.

rabbit *n*: a device used to check the inside diameter of tubulars. It is a solid piece of metal just barely smaller than the inside diameter of the tubular being checked. It is hoisted on a sand line to the derrickhand, who drops it through the drill pipe stand.

rack *n*: 1. framework for supporting or containing a number of loose objects, such as pipe. See *pipe rack*. 2. a bar with teeth on one face for gearing with a pinion or worm gear. 3. a notched bar used as a ratchet. *v*: 1. to place on a rack. 2. to use as a rack.

rack pipe *v*: 1. to place pipe withdrawn from the hole on a pipe rack. 2. to stand pipe on the derrick floor when pulling it out of the hole.

ram blowout preventer *n*: a blowout preventer that uses rams to seal off pressure on a hole that is with or without pipe. Also called a ram preventer. Compare *annular blowout preventer*.

ram preventer *n*: see *ram blowout preventer*.

rathole *n*: a hole in the rig floor, some 30 to 40 feet (9 to 12 metres) deep, which is lined with casing that projects above the floor and into which the kelly and the swivel are placed when hoisting operations are in progress.

RCRA *abbr*: Resource Conservation and Recovery Act.

reeve *v*: to pass (for example, a rope) through a hole or opening in a block or similar device.

reeve the line *v*: to string a wire rope drilling line through the sheaves of the traveling and crown blocks to the hoisting drum.

reeving *n*: a rope system where the rope travels around drums and sheaves.

reservoir pressure *n*: the average pressure within the reservoir at any given time. Determination of this value is best made by bottomhole pressure measurements with adequate shut-in time. If a shut-in period long enough for the reservoir pressure to stabilize is impractical, then various techniques of analysis by pressure buildup or drawdown tests are available to determine static reservoir pressure.

Resource Conservation and Recovery Act (RCRA) *n*: a U.S. federal law that gives the EPA the authority to manage waste at every stage—from its generation and transportation, to its storage and disposal. Under the act, the EPA regulates nonhazardous solid waste, hazardous waste, and underground storage tanks holding hydrocarbon products and some chemicals.

rig *n*: the derrick or mast, drawworks, and attendant surface equipment of a drilling or workover unit.

rig crewmember *n*: see *rotary helper*.

rig down *v*: to dismantle a drilling rig and auxiliary equipment following the completion of drilling operations. Also called tear down.

rig floor *n*: the area immediately around the rotary table and extending to each corner of the derrick or mast—that is, the area immediately above the substructure on which the drawworks and the rotary table rest. Also called derrick floor, drill floor.

rig manager *n*: an employee of a drilling contractor who is in charge of the entire drilling crew and the drilling rig, providing logistics support to the rig crew and liaison with the operating company.

rig operator *n*: see *unit operator*.

rig superintendent *n*: see *toolpusher*.

rig supervisor *n*: see *toolpusher*.

rig up *v*: to prepare the drilling rig for making hole—to install tools and machinery before drilling is started.

roller cone bit *n*: a drilling bit made of two, three, or four cones, or cutters, that are mounted on extremely rugged bearings. The surface of each cone is made of rows of steel teeth or rows of tungsten carbide inserts. Also called rock bit.

rotary *n*: the machine used to impart rotational power to the drill stem while permitting vertical movement of the pipe for rotary drilling. Modern rotary machines have a special component, the rotary or master bushing, to turn the kelly bushing, which permits vertical movement of the kelly while the stem is turning.

rotary bushing *n*: see *master bushing*.

rotary drilling *n*: a drilling method in which a hole is drilled by a rotating bit to which a downward force is applied. The bit is fastened to and rotated by the drill stem, which also provides a passageway through which the drilling fluid is circulated. Additional joints of drill pipe are added as drilling progresses.

rotary helper *n*: a worker on a drilling or workover rig subordinate to the driller, whose primary work station is on the rig floor. On rotary drilling rigs, there are at least two and usually three or more rotary helpers on each crew. Sometimes called floorhand, floorhand, rig crewmember, or roughneck.

rotary hose *n*: a steel-reinforced, flexible hose. It conducts drilling mud from the standpipe to the swivel or top drive. Also called kelly hose or mud hose.

rotary table *n*: the principal piece of equipment in the rotary table assembly; a turning device used to impart rotational power to the drill stem while permitting vertical movement of the pipe for rotary drilling. The master bushing fits inside the opening of the rotary table. It turns the kelly bushing, which permits vertical movement of the kelly while the stem is turning. See *kelly bushing*, *master bushing*.

rotor *n*: 1. a device with vanelike blades attached to a shaft. The device turns or rotates when the vanes are struck by a fluid directed there by a stator. 2. the rotating part of an induction-type alternating current electric motor.

roughneck *n*: see *rotary helper*.

round trip *n*: the procedure of pulling out and subsequently running back into the hole a string of drill pipe or tubing. Also called tripping.

roustabout *n*: 1. a worker on an offshore rig who handles the equipment and supplies that are sent to the rig from the shore base. The head roustabout is very often the crane operator. 2. a worker who assists the superintendent in the general work around a producing oilwell, usually on the property of the oil company. 3. a helper on a well servicing unit.

run casing *v*: to lower a string of casing into the hole. Also called run pipe.

S

safety clamp *n*: a clamp placed very tightly around a drill collar that is suspended in the rotary table by drill collar slips. Should the slips fail, the clamp is too large to go through the opening in the rotary table and therefore prevents the drill collar string from falling into the hole. Also called drill collar clamp.

safety data sheet (SDS) *n*: a document providing information on potential health hazards of chemicals to workers who handle them. As of June 1, 2015 these sheets will take the place of the material safety data sheets (MSDSes) mandated by the Hazard Communication Standard in order to align with the Globally Harmonized System of Classification and Labelling of Chemicals (GHS). See *material safety data sheet, Globally Harmonized System of Classification and Labelling of Chemicals, Hazard Communication Standard*.

safety goggles *n*: a protective eye covering worn to minimize the danger to the eyes of being struck by flying objects or harmed by corrosive substances.

safety joint *n*: an accessory to a fishing tool, placed above it. If the tool cannot be disengaged from the fish, the safety joint permits easy disengagement of the string of pipe above the safety joint. Thus, part of the safety joint and the tool attached to the fish remain in the hole and become part of the fish.

safety latch *n*: a latch provided on a hook to prevent an object suspended from the hook from accidentally slipping or falling out of it.

safety platform *n*: the monkeyboard, or platform on a derrick or mast on which the derrickhand works while wearing a safety harness (attached to the mast or derrick) to prevent falling.

safety shoes *n pl*: metal-toed shoes or boots with nonskid, corrosion-resistant soles worn by oilfield workers to minimize falls and injury to their feet.

safety slide *n*: a wireline device normally mounted near the monkeyboard to afford the derrickhand a means of quick exit to the surface in case of emergency. It is usually affixed to a wireline, one end of which is attached to the derrick or mast and the other end to the surface. To exit by the safety slide, the derrickhand grasps a handle on it and rides it down to the ground. Also called a Geronimo.

safety wire *n*: a steel cable attached to the monkeyboard and anchored to the ground at some distance from the rig. It is used by the derrickhand to slide clear of danger in an emergency.

sandline *n*: a wireline used on drilling rigs and well-servicing rigs to operate a swab or bailer to retrieve cores or to run logging devices. It is usually $\frac{9}{16}$ of an inch (14 millimetres) in diameter and several thousand feet or metres long.

SARA *abbr*: Superfund Amendments and Reauthorization Act.

scaling tool *n*: a circular-shaped wire brush that is attached to a pneumatically, hydraulically, or electrically operated tool (such as a grinder) and that is used to remove rust or scale from pipe or other oilfield equipment.

screen liner *n*: a pipe that is perforated and often arranged with a wire wrapping to act as a sieve to prevent or minimize the entry of sand particles into the wellbore. Also called a screen pipe.

scrubbing *n*: 1. the purification of gas by treatment in a water or a chemical wash. Scrubbing also removes the entrained water in the gas. 2. friction wear.

SDS *abbr*: safety data sheet.

second *n*: 1. the fundamental unit of time in the metric system. 2. in some countries, the crewmember who relieves the toolpusher, or rig manager, and is second in command.

secondary cementing *n*: any cementing operation after the primary cementing operation. Secondary cementing includes a plug-back job, in which a plug of cement is positioned at a specific point in the well and allowed to set. Wells are plugged to shut off bottom water or to reduce the depth of the well for other reasons.

seize *v*: to bind the end of a wire rope with fine wire or a metal band to prevent it from unraveling.

semisubmersible drilling rig *n*: a floating offshore drilling unit that has pontoons and columns that, when flooded, cause the unit to submerge to a predetermined depth. Living quarters, storage space, and so forth are assembled on the deck. Semisubmersible rigs are self-propelled or towed to a drilling site and anchored or dynamically positioned over the site, or both. In shallow water, some semisubmersibles can be ballasted to rest on the seabed. Semisubmersibles are more stable than drillships and barges shaped like ships, so they are used extensively to drill wildcat wells in rough waters such as those of the North Sea. Two types of semisubmersible rigs are the bottle-type and the column-stabilized.

service company *n*: a company that provides a specialized service such as a well-logging service or a directional drilling service.

settling pit *n*: a pit that is dug in the earth for the purpose of receiving mud returned from the well and allowing the solids in the mud to settle out. Steel mud tanks are more often used today, along with various auxiliary equipment for controlling solids quickly and efficiently.

settling tank *n*: 1. the steel mud tank in which solid material in mud is allowed to settle out by gravity. It is used only in special situations today, as solids control equipment has superseded settling tanks, in most cases. Sometimes called a settling pit. 2. a cylindrical vessel on a lease into which produced emulsion is piped and in which water in the emulsion is allowed to settle out of the oil.

shaker *n*: shortened form of shale shaker. See *shale shaker.*

shaker pit *n*: see *shaker tank.*

shaker tank *n*: the mud tank adjacent to the shale shaker, usually the first tank into which mud flows after returning from the hole. Also called a shaker pit.

shale *n*: a fine-grained sedimentary rock composed mostly of consolidated clay or mud. Shale is the most frequently occurring sedimentary rock.

shale shaker *n*: a vibrating screen used to remove cuttings from the circulating fluid in rotary drilling operations. The size of the openings in the screen should be selected carefully to be the smallest size possible to allow 100% flow of the fluid. Also called a shaker.

shear pin *n*: a pin that is inserted at a critical point in a mechanism and designed to shear when subjected to excess stress. The shear pin is usually easy to replace.

shear ram preventer *n*: a blowout preventer that uses shear rams as closing elements.

shear relief valve *n*: a type of pressure relief valve in which excess pressure causes a shearing action on a pin to relieve pressure. See *pressure relief valve.*

sheave (pronounced "shiv") *n*: 1. a grooved pulley. 2. support wheel over which tape, wire, or cable rides.

show *n*: the appearance of oil or gas in cuttings, samples, or cores from a drilling well.

shut-in bottomhole pressure (SIBHP) *n*: the pressure at the bottom of a well when the surface valves on the well are completely closed. It is caused by formation fluids at the bottom of the well.

shut-in bottomhole pressure (SIBHP) test *n*: a bottomhole pressure test that measures pressure after the well has been shut in for a specified period of time. See *bottomhole pressure test.*

sidewall coring *n*: a coring technique that obtains core samples from the hole wall in a zone that has already been drilled. A hollow bullet is fired into the formation wall to capture the core and then retrieved on a flexible steel cable. Core samples of this type usually range from ¾ to 1³⁄₁₆ inches (20 to 30 millimetres) in diameter and from ¾ to 4 inches (20 to 100 millimetres) in length. This method is especially useful in soft-rock areas.

single *n*: a joint of drill pipe. Compare *double, fourble, thribble.*

slips *n pl*: wedge-shaped pieces of metal with serrated inserts (bowls) or other gripping elements, such as serrated buttons, that suspend the drill pipe or drill collars in the master bushing of the rotary table when it is necessary to disconnect the drill stem from the kelly or from the top-drive unit's drive shaft. Rotary slips fit around the drill pipe and wedge against the master bushing to support the pipe. Drill collar slips fit around a drill collar and wedge against the master bushing to support the drill collar. Power slips are pneumatically or hydraulically actuated devices that allow the crew to dispense with the manual handling of slips when making a connection.

snake *n*: see *swivel-connector grip.*

snatch block *n*: 1. a block that can be opened to receive wire rope or wireline. 2. a block that is suited for a single sheave and is used for pulling horizontally on an A-frame mast.

snipe *n*: see *cheater.*

snub *v*: 1. to force pipe or tools into a high-pressure well that has not been killed—to run pipe or tools into the well against pressure when the weight of pipe is not great enough to force the pipe through the BOPs. Snubbing usually requires an array of wireline blocks and a wire rope that forces the pipe or tools into the well through a stripper head or blowout preventer until the weight of the string is sufficient to overcome the lifting effect of the well pressure on the pipe in the preventer. In workover operations, snubbing is usually accomplished by using hydraulic power to force the pipe through the stripping head or blowout preventer. 2. to tie up short with a line.

snubbing line *n*: 1. a line used to check or restrain an object. 2. a wire rope used to put pipe or tools into a well while the well is closed in. See *snub.*

snubbing unit *n*: either a stand-alone device or a rig-assist device that is used to force pipe into the well when the well is shut in on a kick. When the pipe's weight is not sufficient to overcome the upward force of well pressure, a snubbing unit must be used. Compare *stripping in.*

snub line *n*: a strong wire rope attached to the end of the tongs and to one leg of the derrick to keep the tongs from turning too far when they are being used to make up, break out, or back up drill pipe or drill collars.

spinning cathead *n*: see *makeup cathead, spinning chain.*

spinning chain *n*: a Y-shaped chain used to spin up (tighten) one joint of drill pipe into another. In use, one end of the chain is attached to the tongs, another end is attached to the makeup cathead, and the third end is free. The free end is wrapped around the tool joint, and the cathead pulls the chain off the joint, causing the joint to spin (turn) rapidly and tighten up. After the chain is pulled off the joint, the tongs are secured in the same spot, and continued pull on the chain (and thus on the tongs) by the cathead makes up the joint to final tightness.

spud *v*: 1. to move the drill stem up and down in the hole over a short distance without rotation. Careless execution of this operation creates pressure surges that can cause a formation to break down, resulting in lost circulation. 2. to force a wireline tool or tubing down the hole by using a reciprocating motion. 3. to begin drilling a well; to spud in.

spud in *v*: to begin drilling; to start the hole.

squeeze cementing *n*: the forcing of cement slurry by pressure to specified points in a well to cause seals at the squeeze points. It is a secondary cementing method that is used to isolate a producing formation, seal off water, repair casing leaks, and so forth.

stab *v*: to guide the end of a pipe into a tool joint when making up a connection. See *tool joint.*

stabbing board *n*: a temporary platform erected in the derrick or mast some 20 to 40 feet (6 to 12 metres) above the derrick floor. The derrickhand or another crewmember works on the board while casing is being run in a well. The board may be fabricated of wood or of steel girders floored with antiskid material and powered electrically to be raised or lowered to the desired level. A stabbing board serves the same purpose as a monkeyboard but is temporary instead of permanent.

stabilizer *n*: 1. a tool placed on a drill collar near the bit that is used, depending on where it is placed, either to maintain a particular hole angle or to change the angle by controlling the location of the contact point between the hole and the collars. See *packed-hole assembly.* 2. a vessel in which hydrocarbon vapors are separated from liquids. 3. a fractionation system that reduces the vapor pressure so that the resulting liquid is less volatile.

stand *n*: the connected joints of pipe racked in the derrick or mast when a trip is being made. On a rig, the usual stand is about 90 feet (27 metres) long. It is three lengths of drill pipe screwed together, or a thribble.

standard derrick *n*: a derrick that is built piece by piece at the drilling location, as opposed to a jackknife mast, which is preassembled. Standard derricks have been replaced almost totally by jackknife masts. Compare *mast.*

standpipe *n*: a vertical pipe rising along the side of the derrick or mast, which joins the discharge line leading from the mud pump to the rotary hose and through which mud is pumped into the hole.

stator *n*: 1. a device with vanelike blades that serves to direct a flow of fluid (such as drilling mud) onto another set of blades (called the rotor). The stator does not move; rather, it serves merely to guide the flow of fluid at a suitable angle to the rotor blades. 2. the stationary part of an induction-type alternating-current electric motor. Compare *rotor*.

stripper head *n*: a blowout prevention device consisting of a gland and packing arrangement bolted to the wellhead. It is often used to seal the annular space between tubing and casing.

stripping in *n*: 1. the process of lowering the drill stem into the wellbore when the well is shut in on a kick and when the weight of the drill stem is sufficient to overcome the force of well pressure. 2. the process of putting tubing into a well under pressure.

submersible *n*: 1. a two-person submarine used for inspection and testing of offshore pipelines. 2. a submersible drilling rig.

submersible drilling rig *n*: a mobile bottom-supported offshore drilling structure with several compartments that are flooded to cause the structure to submerge and rest on the seafloor. Submersible rigs are designed for use in shallow waters to a maximum of 175 feet (53.4 metres). Submersible drilling rigs include the posted barge submersible, the bottle-type submersible, and the arctic submersible.

substructure *n*: the foundation for the derrick or mast and usually the drawworks. It contains space for storage and well-control equipment.

sucker rod *n*: a special steel pumping rod. Several rods screwed together make up the mechanical link from the beam pumping unit on the surface to the sucker rod pump at the bottom of a well. Sucker rods are threaded on each end and manufactured to dimension standards and metal specifications set by the petroleum industry. Lengths are 25 or 30 feet (7.6 or 9.1 metres); diameters vary from ½ to 1⅛ inches (12 to 30 millimetres). There is also a continuous sucker rod whose trade name is Corod®.

sucker rod pumping *n*: a method of artificial lift in which a subsurface pump located at or near the bottom of the well and connected to a string of sucker rods is used to lift the well fluid to the surface. The weight of the rod string and fluid is counterbalanced by weights attached to a reciprocating beam or to the crank member of a beam pumping unit or by air pressure in a cylinder attached to the beam.

suction pit *n*: also called a suction tank, sump pit, or mud suction pit. See *suction tank*.

suction tank *n*: the mud tank from which mud is picked up by the suction of the mud pumps. Also called a mud suction pit, sump pit, or a suction pit.

Superfund Amendments and Reauthorization Act (SARA) *n*: a U.S. federal law that amended the Comprehensive Response, Compensation, and Liability Act (CERCLA). SARA provided new enforcement authorities, increased the involvement of state governments, and encouraged citizen participation in the process that decides how sites with hazardous waste will be cleaned up. It also increased the size of the trust fund for cleaning up those sites. See Comprehensive Response, Compensation, and Liability Act (CERCLA).

swab *n*: a hollow, rubber-faced cylinder mounted on a hollow mandrel with a pin joint on the upper end to connect to the swab line. A check valve that opens upward on the lower end provides a way to remove the fluid from the well when pressure is insufficient to support flow. *v*: 1. to operate a swab on a wireline to bring well fluids to the surface when the well does not flow naturally. Swabbing is a temporary operation to determine whether the well can be made to flow. If the well does not flow after being swabbed, a pump is installed as a permanent lifting device to bring the oil to the surface. 2. to pull formation fluids into a wellbore by raising the drill stem at a rate that reduces the hydrostatic pressure of the drilling mud below the bit.

swabbing effect *n*: a phenomenon characterized by formation fluids being pulled or swabbed into the wellbore when the drill stem and bit are pulled up the wellbore fast enough to reduce the hydrostatic pressure of the mud below the bit. If enough formation fluid is swabbed into the hole, a kick can result.

swivel *n*: a rotary tool that is hung from the hook and the traveling block to suspend and permit free rotation of the drill stem. It also provides a connection for the rotary hose and a passageway for the flow of drilling fluid into the drill stem.

swivel-connector grip *n*: a braided-wire device used to join the end of one wire rope to the end of another wire rope temporarily. When tension is put on this device, it stretches and grips the wire ropes firmly, allowing the wire rope to be threaded through the blocks. When tension is released, this device relaxes, allowing the rope to be released. Also called a snake or a swivel-type stringing grip.

T

tang *n*: a piece that provides an extension of an instrument (such as a file) and serves to form the handle or make a connection for the attachment of a handle.

thribble *n*: a stand of pipe made up of three joints and handled as a unit. Compare *double, fourble, single*.

thribble board *n*: the name used for the derrickhand's working platform, the monkeyboard, when it is located at a height in the derrick equal to three lengths of pipe joined together. Compare *double board, fourble board*.

tong hand *n*: the member of the drilling crew who handles the tongs.

tongman *n*: see *tong hand*.

tongs *n pl*: the large wrenches used for turning when making up or breaking out drill pipe, casing, tubing, or other pipe; variously called casing tongs, pipe tongs, and so forth, according to the specific use. Power tongs or power wrenches are pneumatically or hydraulically operated tools that serve to spin the pipe up tight and, in some instances, to apply the final makeup torque.

tool hand *n*: a worker on a well service or workover rig who helps run packers and other tools into the well.

toolhouse *n*: a building for storing tools.

tool joint *n*: a heavy coupling element for drill pipe made of special alloy steel. Tool joints have coarse, tapered threads and seating shoulders designed to sustain the weight of the drill stem, withstand the strain of frequent coupling and uncoupling, and provide a leakproof seal. The male section of the joint, or the pin, is attached to one end of a length of drill pipe, and the female section, or the box, is attached to the other end. The tool joint is usually friction welded to the end of the pipe.

toolpusher *n*: an employee of a drilling contractor who is in charge of the entire drilling crew and the drilling rig. Also called a drilling foreman, rig manager, rig superintendent, or rig supervisor.

top drive *n*: a device similar to a power swivel that is used in place of the rotary table to turn the drill stem. It also includes power tongs. Modern top drives combine the elevator, the tongs, the swivel, and the hook. Even though the rotary table assembly is not used to rotate the drill stem and bit, the top-drive system retains it to provide a place to set the slips to suspend the drill stem when drilling stops.

torque *n*: the turning force that is applied to a shaft or other rotary mechanism to cause it to rotate. Torque is measured in foot-pounds, joules, newton-metres, and so forth.

tour (pronounced "tower") *n*: a working shift for drilling crew or other oilfield workers. Some tours are 8 hours; the 3 daily tours are called daylight, evening (or afternoon), and graveyard (or morning). Often, 12-hour tours are used, especially on offshore rigs; they are called simply day tour and night tour.

traveling block *n*: an arrangement of pulleys, or sheaves, through which drilling line is reeved and which moves up and down in the derrick or mast. See *block*.

tubing *n*: relatively small-diameter pipe that is run into a well to serve as a conduit for the passage of oil and gas to the surface.

U

United States Coast Guard (USCG) *n*: an agency of the United States Department of Transportation that establishes regulations for offshore vessels and approves standards for plan approval, construction, operation, and inspection of the hull, machinery, electric systems, industrial systems, and safety and lifesaving equipment of mobile offshore drilling units. The agency conducts rig stability tests, awards the original certification of the rig, and conducts annual inspections. It also establishes manning and personnel licensing requirements and investigates casualties. Address: 2100 Second Street SW; Washington, D.C. 20593; 202-267-2229; www.uscg.mil.

unit operator *n*: the oil company in charge of development and production in an oilfield in which several companies have joined to produce the field. Also called crew chief, rig operator.

USCG *abbr*: United States Coast Guard.

V

valve *n*: a device used to control the rate of flow in a line to open or shut off a line completely, or to serve as an automatic or semiautomatic safety device.

V-**door** *n*: an opening at floor level in a side of a derrick or mast. The V-door is opposite the drawworks and is used as an entry to bring in drill pipe, casing, and other tools from the pipe rack. The name comes from the fact that on the old standard derrick, the shape of the opening was an inverted V.

W

walking beam *n*: the horizontal steel member of a beam pumping unit that has rocking or reciprocating motion.

well completion *n*: 1. the activities and methods of preparing a well for the production of oil and gas or for other purposes, such as injection; the method by which one or more flow paths for hydrocarbons are established between the reservoir and the surface. 2. the system of tubulars, packers, and other tools installed beneath the wellhead in the production casing; that is, the tool assembly that provides the hydrocarbon flow path or paths.

wellhead *n*: the equipment installed at the surface of the wellbore. A wellhead includes such equipment as the casinghead and tubing head. *adj*: pertaining to the wellhead (for example, wellhead pressure).

wickered *adj*: of or having wickers. Also called birdcaged.

wickers *n pl*: loose, broken, frayed, or unraveled strands of the steel wire that makes up the outer wrapping of wire rope.

wildcat *n*: 1. a well drilled in an area where no oil or gas production exists. 2. (nautical) the geared sheave of a windlass used to pull anchor chain. *v*: to drill wildcat wells.

wild well *n*: a well that has blown out of control and from which oil, water, or gas is escaping with great force to the surface. Also called a gusher.

winch *n*: a machine that pulls or hoists by winding a cable around a spool.

windbreak *n*: something that breaks the force of the wind. For example, canvas windbreaks installed around the outside of the rig floor on a drilling or workover rig afford the crew protection from strong, cold winds. Sometimes called a prefab.

wind girder *n*: see *wind ring*.

wind guy line *n*: the wireline attached to ground anchors to provide lateral support for a mast or derrick. Compare *load guy line*.

wind ring *n*: a horizontal stiffening and structural member installed near the top of a floating-roof tank to reinforce the tank wall against wind pressure. Also called a wind girder.

wireline *n*: a slender, rodlike or threadlike piece of metal usually small in diameter or a special current-conducting cable that is used for lowering special tools (such as logging sondes and perforating guns) into the well. Compare *wire rope*.

wire rope *n*: a cable composed of steel wires twisted around a central core of fiber or steel wire to create a rope of great strength and considerable flexibility. Wire rope is used as drilling line (in rotary and cable-tool rigs), coring line, servicing line, winch line, and so on. It is often called cable or wireline; however, wireline is a single, slender metal rod that is usually very flexible. Compare *wireline*.

workover rig *n*: a portable rig used for working over a well. See *production rig*.

wrist pin *n*: in an engine, the hollow cylinder of hard steel that attaches the piston to the piston rod. A circular opening in the piston is lined up with a corresponding circular opening in the rod and the wrist pin is pushed through the openings. Usually, special keys on each end of the pin lock the pin in the piston. Also called piston pin.

Review Questions
LESSONS IN ROTARY DRILLING
Unit I, Lesson 10: Safety on the Rig

Fill in the Blanks

Fill in the blanks with the appropriate word or phrase.

1. Drilling rig safety is the concern of _______________________________________
   ____ .

2. List the four leading causes of drilling rig accidents:

   (a) ____

   (b) ____

   (c) ____

   (d) ____

3. Name four items of personal protective equipment (PPE) worn or used by drilling rig personnel:

   (a) ____

   (b) ____

   (c) ____

   (d) ____

4. Avoidable accidents happen during rig-up operations because _________________
   ____ .

5. Strains are the leading cause of back injuries. When lifting a heavy or awkward load, you should
   always ____ .

6. The primary rule for hand tool safety is _________________________________ .

7. For transportation offshore you must _______________________________ and be aware that
   the _________________ or _________________ is in complete control during the trip.

8. A_________________ and _________________________ should be provided for each employee
   who works above the first girt of a derrick or mast.

9. Only the _________________ or other _________________ personnel are allowed to operate
   the drawworks or rotary. However, _________________ should be instructed in how to
   shut them down in the event of an emergency.

10. Chain hoists or snatch blocks should not be fastened to derrick girts because any ______________________ of a derrick member can ______________ the derrick.

11. A clear, smooth working area is necessary around the rotary. The main reason to keep all tools picked up on the derrick floor is ____ .

    (a) to keep from losing them.

    (b) to prevent tripping on them.

    (c) to keep the tools clean.

12. The rotary clutch should not be engaged if either drill pipe tong is latched onto the pipe. Exceptions are:

    (a) ____

    (b) ____

13. Crewmembers should not stand ______________________________ the two tongs while pipe is being made up or broken out.

14. No one should operate a spinning chain without first being ____________________ in its use. The chain should not be permitted to become lodged in the thread of a tool joint because the chain may dent or bruise the ______________ and ______________ shoulder when they are drawn together.

15. Three safety rules to observe when operating a catline and cathead include—

    (a) Never ____________________________ than are necessary on the cathead.

    (b) Never work on the cathead if no one is at the ____________________ .

    (c) Never stand within the ____________________ on the floor.

16. Moving pipe deserves your full attention. Rule #1 around the drilling rig is:

    ____ .

17. Three important safety rules for working with electricity include the following:

    (a) Equipment repairs, cutting, or splicing of electrical wiring should only be done by ______________ .

    (b) Avoid contact with any ____________________ .

    (c) ______________ and electricity do not mix; therefore, never use ______________ to extinguish an electrical fire and do not operate electrical equipment where ______________ hazards are present.

18. Name two types of activities where regulations require a permit before work begins:

    (a) ____

    (b) ____

19. Name two rules for chemical and caustic handling:

    (a) ____

    (b) ____

20. Time is crucial in controlling a well kick. Early recognition of the warning signs is critical. List five signs that a well kick is imminent:

    (a) ____

    (b) ____

    (c) ____

    (d) ____

    (e) ____

21. After a well kick happens, what two procedures are required to regain a full column of mud heavy enough to contain the formation pressure?

    (a) ____

    ____

    (b) ____

    ____

22. List the four classes of fires and the type fuel for each:

    (a) ____

    (b) ____

    (c) ____

    (d) ____

23. List the three types of fire extinguishers and the classes of fire for which each is used:

    (a) ____

    (b) ____

    (c) ____

24. Every rig should have standard procedures to follow in the event of an H_2S emergency. List three recommended actions by crewmembers:

    (a) ____

    (b) ____

    (c) ____

25. A fellow crewmember has been seriously injured in a rig accident. What are the three basic steps you should take? List them in order.

    (a) ____

    (b) ____

    (c) ____

Index

accidents. See also *lost-time injuries (LTIs)*
   factors increasing risk of, 39
accumulator units, 101
adjustable wrenches, 27
"Adult First Aid/CPR/AED," 148
Air Force, 121
air hoist, 66–67
air-powered kelly spinner, 50
air-powered spinning tongs, 59
air supply equipment, 136–137
air tools, safety devices on, 34
alarm boxes, manual, 133
alcohol
   offshore installations and, 20
   use, 8, 9, 21, 22, 24
alertness, 5, 39
Allen wrenches, 27, 75
American Red Cross Manual, 148
anchor drum, 53
anti-crowning device, 51
antifouling device, 62
arcing, electrical equipment subject to, 92
artificial resuscitation. See also *rescue breathing*
electrical shock and, 90
atmosphere, hazardous, 94
automatic cathead, 63
automatic sprinklers, 130

bails, 61
basket-lift transfer, 23
bell nipple, 139
blade switch, 88
bleeding, 149
blowout(s)
   early warning signs of, 101
   fatalities and, 19
   risk of, 108
blowout preventer (BOP)
   about, 99, 100
   DST tool and, 107
   installation and, 100–101
   safety rules for, 104

blowout preventer stack, 100
bone fractures, 159–160. See also *head injuries; spine injuries*
BOP. See *blowout preventer (BOP)*
borehole, 108
bottleneck elevators, 61
brake handle, 45
brakes, drawworks and, 47
breaker switches, 87, 88, 92
breakout, 58
breathing equipment
   air supply, 136–137
   categories of, 144
   H_2S gas and, 142, 143, 144
   perforated eardrum and, 140
   positive-pressure air masks, 144
bridle, 109
brushes, 30
BSEE. See *Bureau of Safety and Environmental Enforcement (BSEE)*
buddy system
   confined spaces and, 94
   H_2S gas and, 142
bulbs, 89
bumper blocks, 51
bunker suit
   air supply, 136–137
   coat and pants, 134, 135
   components of, 134
   maintenance of, 136
Bureau of Safety and Environmental Enforcement (BSEE), 19
burn(s)
   about, 151
   chemical, 153
   electrical, 153
   first-degree, 151
   second-degree, 151
   third-degree, 151
   treatment of, 152–153
burn pit, 116
butane, 117

Answers to Review Questions
LESSONS IN ROTARY DRILLING
Unit I, Lesson 10: Safety on the Rig

1. everyone having anything to do with the job.

2. a. Highway accidents
   b. Being struck by an object
   c. Explosions
   d. Being caught in moving machinery

3. a. Safety hat
   b. Safety shoes
   c. Goggles or face shields
   d. Gloves
   e. Respirators
   f. Safety harness
   g. Hearing protection
   h. Chemical apron and gloves

4. the work is nonroutine, heavy machinery and equipment are moving about, and many jobs go on at the same time in close quarters

5. get help or use hoisting equipment

6. use the right tool for the job

7. report on time …. pilot or captain

8. safety harness and lifeline

9. driller …. authorized/designated …. everyone

10. bending ….weaken

11. b

12. a. When tongs are used to hold a joint or stand of pipe after the connection has been broken or loosened with two tongs
    b. When backup tongs are so tight they cannot be released after a joint has been broken unless the rotary is turned slightly

13. between or within the arc of

14. instructed …. pin and box

15. a. add more wraps
    b. driller's controls
    c. coils of rope

16. Never turn your back on moving pipe.

17. a. authorized personnel
    b. exposed wiring
    c. Water….water….water

18. a. welding/cutting
    b. work in confined spaces

19. a. Never remove labels without replacing them.
    b. Wear personal protective equipment.
    c. Read the safety data sheet (SDS).
    d. Never assume an unlabeled container is harmless.
    e. Never mix unlabeled chemicals.

20. a. Drilling break
    b. Flow of fluid from the well
    c. Rising pit level
    d. Decrease in circulation pressure
    e. Shows of oil, gas, or saltwater in the drilling mud

21. a. Close the well in using blowout preventers and chokes.
    b. Circulate out the extraneous fluid that has entered the wellbore by using an adjustable choke and reduced circulating pressure from the mud pumps.

22. a. Class A—common materials like paper, wood, trash, and rags
    b. Class B—gasoline, oil, paint thinner, and grease
    c. Class C—electrical wiring and equipment
    d. Class D—combustible metals such as magnesium, sodium, lithium, titanium, and alloys

23. a. Carbon dioxide—Class B and C
    b. Dry chemical—Class A, B, C
    c. Dry powder—Class D

24. a. Hold breath and put on breathing equipment, if it is readily available.

b. Move in an upwind direction away from the H$_2$S source.

c. Assemble in a designated safe area.

25. a. CHECK the area and victims.

b. CALL for emergency medical help.

c. CARE for the victims.